Everyday Mathematics®

The University of Chicago School Mathematics Project

Student Math Journal
Volume 1

Grade 2

McGraw Hill Education

Chicago, IL • Columbus, OH • New York, NY

The University of Chicago School Mathematics Project (UCSMP)

Max Bell, Director, UCSMP Elementary Materials Component; Director, *Everyday Mathematics* First Edition; James McBride, Director, *Everyday Mathematics* Second Edition; Andy Isaacs, Director, *Everyday Mathematics* Third Edition; Amy Dillard, Associate Director, *Everyday Mathematics* Third Edition; Rachel Malpass McCall, Associate Director, *Everyday Mathematics* Common Core State Standards Edition

Authors

Max Bell, Andy Isaacs, Jean Bell, James McBride, John Bretzlauf, Cheryl G. Moran, Amy Dillard, Kathleen Pitvorec, Robert Hartfield, Peter Saecker

Technical Art
Diana Barrie

Third Edition Teachers in Residence
Kathleen Clark, Patti Satz

UCSMP Editorial
John Wray, Don Reneau

Contributors

Mikhail Guzowski, Robert Balfanz, Judith Busse, Mary Ellen Dairyko, Lynn Evans, James Flanders, Dorothy Freedman, Nancy Guile Goodsell, Pam Guastafeste, Nancy Hanvey, Murray Hozinsky, Deborah Arron Leslie, Sue Lindsley, Mariana Mardrus, Carol Montag, Elizabeth Moore, Kate Morrison, William D. Pattison, Joan Pederson, Brenda Penix, June Ploen, Herb Price, Dannette Riehle, Ellen Ryan, Marie Schilling, Susan Sherrill, Patricia Smith, Robert Strang, Jaronda Strong, Kevin Sweeney, Sally Vongsathorn, Esther Weiss, Francine Williams, Michael Wilson, Izaak Wirzup

Photo Credits

Cover (l)Linda Lewis/Frank Lane Picture Agency/CORBIS, (r)C Squared Studios/Getty Images, (bkgd)Estelle Klawitter/Cusp/CORBIS; **Back Cover Spine** C Squared Studios/Getty Images; **iii-vi** The McGraw-Hill Companies; **vii** (t)Siede Preis/Photodisc/Getty Images, (b)Stockbyte; **viii** Burke/Triolo Productions/Getty Images; **2-54** The McGraw-Hill Companies; **93** (br)Brand X Pictures/PunchStock, (others)The McGraw-Hill Companies; **159** The McGraw-Hill Companies.

everyday**math**.com

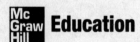

STEM McGraw-Hill is committed to providing instructional materials in Science, Technology, Engineering, and Mathematics (STEM) that give all students a solid foundation, one that prepares them for college and careers in the 21st century.

Send all inquiries to:
McGraw-Hill Education
STEM Learning Solutions Center
P.O. Box 812960
Chicago, IL 60681

ISBN: 978-0-07-657634-0
MHID: 0-07-657634-5

Printed in the United States of America.

10 11 12 13 QVS 17 16 15 14

The **McGraw·Hill** Companies

Contents

UNIT 1 Numbers and Routines

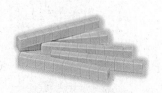

Contents **v**

UNIT 6 Whole-Number Operations and Number Stories

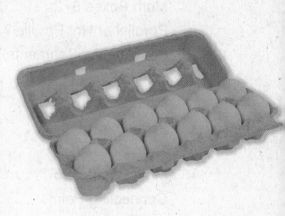

Projects

Activity Sheets

LESSON 1·1   # Number Sequences

Fill in the missing numbers.

1.
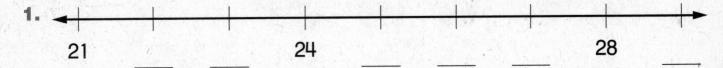

21 ___ ___ 24 ___ ___ ___ 28 ___

2.
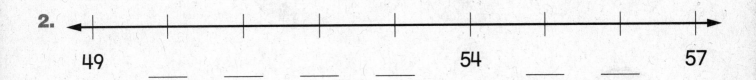

49 ___ ___ ___ ___ 54 ___ ___ 57

3.
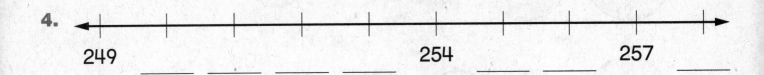

127 ___ ___ 130 ___ ___ 134 ___ ___

4.
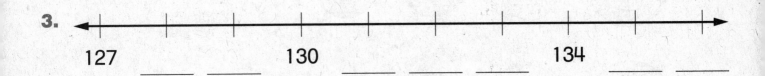

249 ___ ___ ___ ___ 254 ___ ___ 257 ___

5.
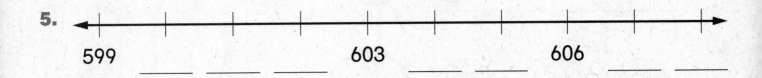

599 ___ ___ ___ 603 ___ ___ 606 ___

6.
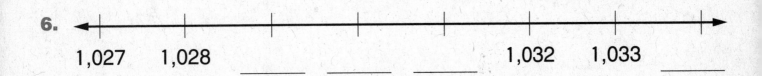

1,027   1,028 ___ ___ ___ ___ 1,032   1,033 ___

LESSON 1·2 Coins

1.   = _____ ¢

2.       = _____ ¢

3.    = _____ ¢

4.    = _____ ¢

5.      = _____ ¢

6.    = _____ ¢

LESSON 1·3

Calendar for the Month

Month _____

Sunday	Monday	Tuesday	Wednesday	Thursday	Friday	Saturday

LESSON 1·3   **Time**

Write the time.

1.

2.

3.

_____ : _____   _____ : _____   _____ : _____

Draw the hands.

4.

5.

6.

5:00   9:30   1:45

Draw the hands and write the time.

7.

8.

9.

_____ : _____   _____ : _____   _____ : _____

LESSON 1·4 **Addition Facts** 5

1.

   2 + 4 = _____

2.

   5 + 5 = _____

3.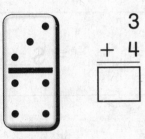

   $\begin{array}{r} 3 \\ + 4 \\ \hline \square \end{array}$

4.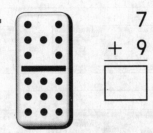

   $\begin{array}{r} 7 \\ + 9 \\ \hline \square \end{array}$

5. 6 + 8 = _____   6. 8 + _____ = 11   7. _____ + 4 = 10

___

Addition Top-It

Write a number model for each player's cards.

Then write <, >, or = in the box.

_____ = _____   _____ = _____

_____ = _____   _____ = _____

_____ = _____   _____ = _____

LESSON 1·5

Counting Bills

Write the amount.

1.

= $_____

2.

= $_____

3.

= $_____

4.

= $_____

Try This

5.

= $_____

LESSON 1·6   Math Boxes

1. Fill in the missing numbers.

a. 36, _____, 38, _____

b. _____, 12, 13, _____

c. 89, _____, 91, _____

d. _____, 147, _____, 149, _____

2. Circle the tens digit.

437

Circle the ones digit.

18

Circle the hundreds digit.

206

3. How likely is it that our class will go on a field trip today? Circle.

certain

likely

unlikely

impossible

4. Today is

_____ _____, _____.
(month)   (day)   (year)

The date 1 week from today will

be _____.

5. Fill in the circle next to the name of the shape.

Ⓐ triangle

Ⓑ rectangle

Ⓒ pentagon

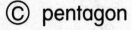

6. Write two even and two odd numbers.

even _____

even _____

odd _____

odd _____

LESSON 1·7   Math Boxes

1. How much money?

_____ ¢

MRB 88 89

2. Fill in the missing numbers.

	4			7
13		15		

MRB 8

3. Show 16 with tally marks.

_____

MRB 40

4. Write the time.

____ : ____

MRB 81

5. Solve.

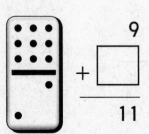

$$+ \boxed{} \begin{array}{r} 9 \\ \hline 11 \end{array}$$

6. Fill in the missing frames.

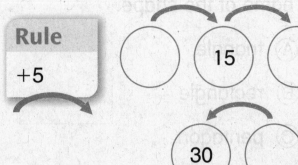

Rule
+5

15

30

MRB 98

8 eight

LESSON 1·8 Number-Grid Puzzles

							80		
9									
7									
						66			
		35							
2									
		21				61			

LESSON 1·8   **Math Boxes**

---

1. Fill in the missing numbers.

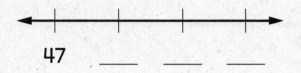

47   _____ _____ _____

2. Write the missing numbers.

124, _____

126, _____

128, _____

---

3. Show 2 ways to make 5 cents.

MRB
88 89

4. Write odd or even.

3   _____

8   _____

1   _____

27 _____

MRB
97

---

5. Fill in the missing frames.

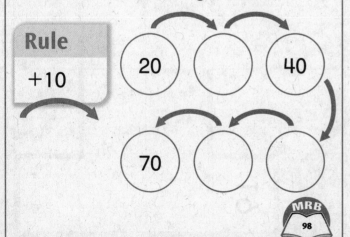

Rule
+10

MRB
98

6. Solve.

Unit

_____ = 6 + 5

7 + 5 = _____

_____ = 8 + 5

9 + 5 = _____

Broken Calculator

Example: Show 17.
Broken key is (7).
Show several ways:

$$11 + 6$$
$$20 - 3$$
$$8 + 8 + 1$$

1. Show 3.
Broken key is (3).
Show several ways:

_____

_____

_____

2. Show 20.
Broken key is (2).
Show several ways:

_____

_____

3. Show 22.
Broken key is (2).
Show several ways:

_____

_____

4. Show 12.
Broken key is (1).
Show several ways:

_____

_____

_____

5. Make up your own.
Show _____.
Broken key is _____.
Show several ways:

_____

_____

_____

LESSON 1·9   Hundreds-Tens-and-Ones Problems

Solve. Use your tool-kit bills to help you.

Example:

2 [$10]   3 [$1]   How much? $_____

1.

8 [$10]   4 [$1]   How much? $_____

2.

5 [$100]   4 [$10]   3 [$1]   How much? $_____

3.

7 [$100]   6 [$10]   9 [$1]   How much? $_____

4.

8 [$100]   2 [$10]   0 [$1]   How much? $_____

5.

3 [$100]   0 [$10]   4 [$1]   How much? $_____

Try This

6.

22 [$100]   5 [$10]   7 [$1]   How much? $_____

LESSON 1·9   **Math Boxes**

---

1. How much money?

Ⓓ Ⓓ Ⓓ Ⓓ Ⓓ Ⓓ

_____ ¢

MRB
88 89

---

2. Fill in the missing numbers.

2			
	13		15

MRB
8

---

3. Show 23 with tally marks.

_____

MRB
40

---

4. What time is it?

_____ : _____

In $\frac{1}{2}$ hour, it will be _____ : _____.

MRB
81

---

5. Solve. Fill in the circle next to the best answer.

Ⓐ 14

Ⓑ 2

_____ + 6 = 8

Ⓒ 8

Ⓓ 10

---

6. Write the arrow rule. Fill in the missing frames.

Rule

5   ◯   15

◯   25   ◯

MRB
98 99

LESSON 1·10 Counting with a Calculator

1. Count by 7s on your calculator. Write the numbers.

   7, _____, _____, _____, _____, _____, _____

   What number is added each time you press ⊜? _____

---

2. Count by 6s on your calculator. Write the numbers.
   Circle the 1s digit in each number.

   6, _____, _____, _____, _____, _____, _____, _____, _____, _____

   What number is added each time you press ⊜? _____

   What pattern do you see in the 1s digits? _____

---

3. Count by 4s on your calculator. Write the numbers.
   Circle the 1s digit in each number.

   4, _____, _____, _____, _____, _____, _____, _____, _____, _____

   What number is added each time you press ⊜? _____

   What pattern do you see in the 1s digits? _____

Try This

4. Jim counted on the calculator. He wrote these numbers: 3, 5, 7, 9, 11.

   What keys did he press? _____

LESSON 1·10   Broken Calculator

1. Show 8.
Broken key is ⑧.
Show several ways:

_____

_____

_____

2. Show 30.
Broken key is ③.
Show several ways:

_____

_____

_____

3. Show 15.
Broken key is ⑤.
Show several ways:

_____

_____

_____

4. Show 26.
Broken key is ⑥.
Show several ways:

_____

_____

_____

5. Make up your own.
Show _____.
Broken key is _____.
Show several ways:

_____

_____

6. Make up your own.
Show _____.
Broken key is _____.
Show several ways:

_____

_____

LESSON 1·10   **Math Boxes**

---

1. Fill in the missing numbers.

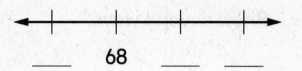

_____   68   _____   _____

---

2. Write these numbers in order. Start with the smallest number.

40   23   81

_____ , _____ , _____

---

3. Show two ways to make 35¢.

---

4. Write even or odd.

20   _____

5   _____

17   _____

33   _____

---

5. Fill in the missing frames.

Rule
+10

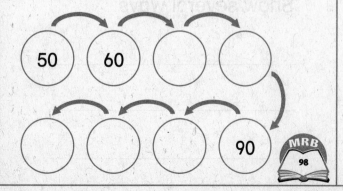

50   60   ( )   ( )

( )   ( )   ( )   90

---

6. SOLVE.

Unit

```
  9      2      6
+ 5    + 5    + 5
[ ]    [ ]    [ ]

  8      5      5
+ 5    + 7    + 5
[ ]    [ ]    [ ]
```

LESSON 1·11 Using <, >, and =

3 < 5 3 is less than 5.	5 > 3 5 is greater than 3.

Write <, >, or =.

1. 61 _____ 26

2. 18 _____ 81

3. 107 _____ 57

4. 114 _____ 114

5. 299 _____ 302

6. 1,032 _____ 1,132

Try This

7. 15 _____ 7 + 8

8. 9 + 2 _____ 4 + 5

9. 5 + 6 _____ 8 + 4

Write the total amounts. Then write <, >, or =.

Example: Ⓓ Ⓝ Ⓝ Ⓟ Ⓟ = __22__ ¢ __<__ __26__ ¢ = Ⓠ Ⓟ

10. Ⓝ Ⓝ Ⓓ Ⓟ = _____ ¢ _____ _____ ¢ = Ⓠ Ⓝ Ⓓ Ⓝ

11. Ⓝ Ⓓ Ⓟ Ⓠ = _____ ¢ _____ _____ ¢ = Ⓝ Ⓓ Ⓓ Ⓝ Ⓟ Ⓠ

12. Ⓓ Ⓓ Ⓠ Ⓓ = _____ ¢ _____ _____ ¢ = Ⓓ Ⓝ Ⓟ Ⓓ

LESSON 1·11 Math Boxes

1. How much money? Circle the best answer.

N P

_____ ¢

A 41¢

B 36¢

C 82¢

D 5¢

MRB 88 89

2. Fill in the missing numbers.

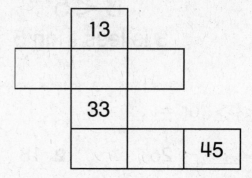

MRB 8

3. Write the number.

|||| |||| |||| ||||
|||| |||| |||

_____

MRB 40

4. Draw hands to show 6:15.

MRB 81

5. Solve.

Unit

$6 + 6 =$ _____

$9 + 9 =$ _____

$20 + 20 =$ _____

_____ $= 100 + 100$

6. Solve.

Unit

$139 + 1 =$ _____

$1 + 479 =$ _____

_____ $= 299 + 1$

_____ $= 398 + 2$

LESSON 1·12 — Math Boxes

1. Fill in the missing numbers.

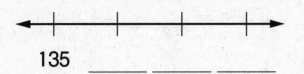

135 _____ _____ _____

2. Fill in the oval next to the numbers that are in order from the smallest to the largest.

◯ 103, 29, 86

◯ 29, 86, 103

◯ 29, 103, 86

◯ 86, 29, 103

3. Write the amount.

Ⓠ Ⓓ Ⓝ Ⓝ Ⓝ Ⓟ

_____¢

4. Write even or odd. You may use counters.

4 _____

7 _____

10 _____

15 _____

5. Write the arrow rule. Fill in the missing frames.

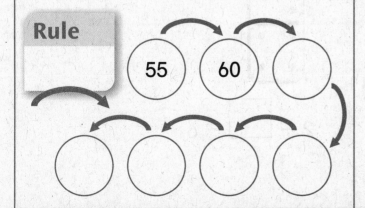

Rule

55 60

6. Fill in the blanks.

25, 35, _____, _____, _____

LESSON 1·13 Math Boxes

1. Write 6 names for 9.

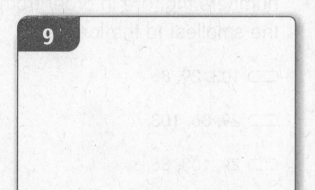

9

2. Solve.

Unit

$4 + 5 =$ _____

$5 + 3 =$ _____

$$\begin{array}{cc} 8 & 11 \\ +\ 5 & +\ 5 \end{array}$$

3. Use a number grid. How many spaces from:

17 to 26? _____

49 to 65? _____

4. Solve.

Unit

$14 + 1 =$ _____

$136 + 1 =$ _____

$291 + 1 =$ _____

$279 + 1 =$ _____

5. Fill in the missing frames.

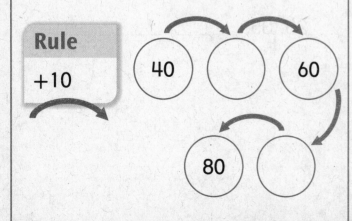

Rule

+10

40   60

80

6. Solve.

$2 +$ ☐ $= 6$

LESSON 2·1   **Number Stories**

Write an addition number story about what you see in the picture. Write a label in the unit box. Find the answer. Write a number model.

Example: *7 ducks in the water. 5 ducks in the grass. How many ducks in all?*

Answer the question: ___*12 ducks*___
                          (unit)

Number model: ___*7*___ + ___*5*___ = ___*12*___

Unit

ducks

Story: _____

_____

_____

_____

_____

Unit

Answer the question: _____
                        (unit)

Number model: _____ + _____ = _____

Date Time

LESSON 2·1

Number-Grid Puzzles

Complete the number-grid puzzles.

11				15					20
							28		
		33							
	52						58		

	332				336				
								349	
						357			
361									
		373					378		

LESSON 2·1

Math Boxes

---

1. Six apples are red. Five apples are green. How many apples in all?

   _____

   Number Model

   Unit

   apples

   _____

2. Use your calculator.

   Show 14.
   Broken key is ①.
   Show 2 ways:

   _____

   _____

---

3. Fill in the blanks.

   83, _____, 81, _____, _____, 78

4. Use < or >.

   4 + 5 _____ 10

   12 _____ 7 + 4

   15 _____ 8 + _____

   6 + 7 _____ 15 − 4

---

5. Write the time.

   _____ : _____ _____

6. How much money? Fill in the circle next to the best answer.

   Ⓐ $2.40   Ⓑ $11.45

   Ⓒ $11.40   Ⓓ $2.45

LESSON 2·2 Fact Power Table

0 +0	0 +1	0 +2	0 +3	0 +4	0 +5	0 +6	0 +7	0 +8	0 +9
1 +0	1 +1	1 +2	1 +3	1 +4	1 +5	1 +6	1 +7	1 +8	1 +9
2 +0	2 +1	2 +2	2 +3	2 +4	2 +5	2 +6	2 +7	2 +8	2 +9
3 +0	3 +1	3 +2	3 +3	3 +4	3 +5	3 +6	3 +7	3 +8	3 +9
4 +0	4 +1	4 +2	4 +3	4 +4	4 +5	4 +6	4 +7	4 +8	4 +9
5 +0	5 +1	5 +2	5 +3	5 +4	5 +5	5 +6	5 +7	5 +8	5 +9
6 +0	6 +1	6 +2	6 +3	6 +4	6 +5	6 +6	6 +7	6 +8	6 +9
7 +0	7 +1	7 +2	7 +3	7 +4	7 +5	7 +6	7 +7	7 +8	7 +9
8 +0	8 +1	8 +2	8 +3	8 +4	8 +5	8 +6	8 +7	8 +8	8 +9
9 +0	9 +1	9 +2	9 +3	9 +4	9 +5	9 +6	9 +7	9 +8	9 +9

Distances on a Number Grid

Example: *How many spaces do you move to go from 17 to 23 on the number grid?*

Solution: *Place a marker on 17. You move the marker 6 spaces before landing on 23.*

11	12	13	14	15	16	17	18	19	20
21	22	23	24	25	26	27	28	29	30

How many spaces from:

23 to 28? _____   15 to 55? _____   39 to 59? _____

27 to 42? _____   34 to 26? _____   54 to 42? _____

15 to 25? _____   26 to 34? _____

1	2	3	4	5	6	7	8	9	10
11	12	13	14	15	16	17	18	19	20
21	22	23	24	25	26	27	28	29	30
31	32	33	34	35	36	37	38	39	40
41	42	43	44	45	46	47	48	49	50
51	52	53	54	55	56	57	58	59	60

LESSON 2·2   **Math Boxes**

1. Count by 3s.
Use your calculator.

_____, $\dfrac{6}{}$, _____, _____,

$\dfrac{15}{}$, _____

MRB
162 163

2. Fill in the missing numbers.

455, _____, _____, 458

3. Solve.

$4 + 3 =$ ____

$10 - 7 =$ ____

```
  5        8
+ 4      - 3
```

Unit

4. Show $1.00 three ways.
Use Ⓠ, Ⓓ, and Ⓝ.

MRB
88–90

5. Mrs. Satz's Class's
Favorite Colors

Colors	Tallies
Red	⊞⊞
Blue	⊞ ///
Green	⊞
Yellow	//

Which color is the
most popular? _____

6. Fill in the circle that names
the number.

5 ones

6 hundreds

3 tens

Ⓐ 564

Ⓑ 356

Ⓒ 635

Ⓓ 536

MRB
10

LESSON 2·3   **Addition/Subtraction Facts Table**

+,−	0	1	2	3	4	5	6	7	8	9
0	0	1	2	3	4	5	6	7	8	9
1	1	2	3	4	5	6	7	8	9	10
2	2	3	4	5	6	7	8	9	10	11
3	3	4	5	6	7	8	9	10	11	12
4	4	5	6	7	8	9	10	11	12	13
5	5	6	7	8	9	10	11	12	13	14
6	6	7	8	9	10	11	12	13	14	15
7	7	8	9	10	11	12	13	14	15	16
8	8	9	10	11	12	13	14	15	16	17
9	9	10	11	12	13	14	15	16	17	18

Domino-Dot Patterns

Draw the missing dots on the dominoes. Find the total number
on both halves.

1. double 2

$\begin{array}{r} 2 \\ +\,2 \\ \hline \end{array}$

2. double 3

$\begin{array}{r} 3 \\ +\,3 \\ \hline \end{array}$

3. double 4

$\begin{array}{r} 4 \\ +\,4 \\ \hline \end{array}$

4. double 5

$\begin{array}{r} 5 \\ +\,5 \\ \hline \end{array}$

5. double 6

$\begin{array}{r} 6 \\ +\,6 \\ \hline \end{array}$

6. double 7

$\begin{array}{r} 7 \\ +\,7 \\ \hline \end{array}$

7. double 8

$\begin{array}{r} 8 \\ +\,8 \\ \hline \end{array}$

8. double 9

$\begin{array}{r} 9 \\ +\,9 \\ \hline \end{array}$

Find the total number of dots.

9.

$\begin{array}{r} 4 \\ +\,5 \\ \hline \end{array}$

10.

$\begin{array}{r} 6 \\ +\,5 \\ \hline \end{array}$

11.

$\begin{array}{r} 6 \\ +\,7 \\ \hline \end{array}$

12.

$\begin{array}{r} 8 \\ +\,7 \\ \hline \end{array}$

LESSON 2·3 *Doubles or Nothing* **Record Sheet**

Round 1

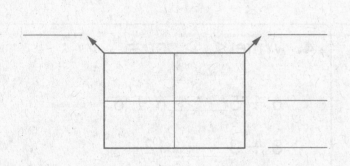

Total _____

Round 2

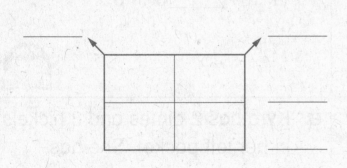

Total _____

Round 3

Total _____

Grand Total _____

LESSON 2·3   **Math Boxes**

---

1. Julie had 10 crayons. Rosa gave her 8 more crayons. How many crayons in all?

_____ crayons

Number model:

_____

Unit
crayons

2. Use your calculator.

Show 25.
Broken key is ⑤.
Show 2 ways:

_____

_____

---

3. Count back by 5s.

45, 40, _____, _____, _____,

_____, _____, 10, _____

Can you keep going?

0, _____, _____

4. Write <, >, or =.

$6 + 5$ _____ $6 + 6$

$8 + 3$ _____ 12

$9 + 9$ _____ 17

$4 + 9$ _____ $8 + 5$

---

5. Draw the hands to show 10:30.

6. Kyra has 2 dimes and 3 nickels in her left pocket. She has 1 quarter and 2 pennies in her right pocket. How much money does she have?

_____

LESSON 2·4 **+9 Facts**

Write the sums.

1. 3
   + 9
   ____

2. 7
   + 9
   ____

3. ____ = 9 + 5

Unit

4. ____ = 2 + 9

5. 9
   + 4
   ____

6. 6
   + 9
   ____

7. 9
   + 8
   ____

8. 1 + 9 = ____

9. 0 + 9 = ____

10. 9
   + 9
   ____

11. 9 + ____ = 12

12. 15 = 9 + ____

13. ____ + 9 = 17

14. 13 = ____ + 9

Unit

Write a +9 number story.

_____

_____

_____

LESSON 2·4 Math Boxes

---

1. Count by 6s.
Use your calculator.

_____, 12, _____, _____, 30

MRB
162 163

---

2. Fill in the missing numbers.

196	
	207

MRB
97

---

3. Solve.

Unit

$16 = $ _____ $+ 1$

_____ $= 14 + 0$

_____ $+ 9 = 9$

_____ $+ 5 = 12$

---

4. Show $0.88 in two ways.
Use Ⓠ, Ⓓ, Ⓝ, and Ⓟ.

MRB
88

---

5. Room 10's Favorite Seasons

Season	Number of Children
Fall	⊬⊬
Winter	⊬⊬ //
Spring	⊬⊬
Summer	⊬⊬ ////

Which seasons have the
same number of votes?

_____

---

6. 132 has… _____ hundreds

_____ tens

_____ ones

MRB
10

LESSON 2·5 — Addition Facts

If you know a double, you know the 1-more and the 1-less sums.

Example:
If you know that $4 + 4 = 8$,
You know $\quad 4 + 5 = 9$,
And $\quad\quad 4 + 3 = 7$

1. $3 + 4 =$ _____   2. $8 + 7 =$ _____

3. _____ $= 6 + 7$ 4. $\begin{array}{r} 6 \\ + 5 \\ \hline \end{array}$

5. $\begin{array}{r} 8 \\ + 9 \\ \hline \end{array}$ 6. $\begin{array}{r} 7 \\ + 5 \\ \hline \end{array}$ 7. $\begin{array}{r} 7 \\ + 9 \\ \hline \end{array}$

8. $5 + 8 =$ _____   9. _____ $= 6 + 9$ 10. $8 + 6 =$ _____

Try This

11. $8 + 8 =$ _____   12. $12 + 12 =$ _____

$\quad\,\, 8 + 9 =$ _____ $\quad\,\, 12 + 13 =$ _____

$\quad\,\, 8 + 7 =$ _____ $\quad\,\, 12 + 11 =$ _____

13. $15 + 15 =$ _____   14. $14 + 12 =$ _____

$\quad\, 16 + 15 =$ _____ $\quad\, 15 + 13 =$ _____

$\quad\, 14 + 15 =$ _____

LESSON 2·5 **Math Boxes**

---

1. What shape is the cover of your math journal? Fill in the circle next to the best answer.

Ⓐ rhombus

Ⓑ rectangle

Ⓒ triangle

Ⓓ square

MRB 52

2. Circle the activity that takes about 1 second.

Blinking your eyes.

Writing your name.

Reading a story.

---

3. Complete the fact family. Fill in the missing domino dots.

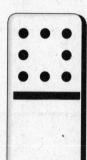

$8 + \underline{\quad} = 15$

$\underline{\quad} + 8 = 15$

$\underline{\quad} = 15 - 8$

$8 = 15 - \underline{\quad}$

MRB 25

4. How likely is it that our school will have a fire drill today? Circle your answer.

certain

likely

unlikely

impossible

---

5. Put the numbers in order from smallest to largest. Circle the middle number.

48, 44, 37, 54, 39

_____ , _____ , _____ , _____ , _____

MRB 46

6. Draw hands to show 8:15.

LESSON 2·6   **Domino Facts**   35

For Problems 1 through 7, write 2 addition facts and 2 subtraction facts for each domino.

1.

$$\begin{array}{c} +\dfrac{\begin{array}{c}4\\2\end{array}}{6} \quad +\dfrac{\begin{array}{c}2\\4\end{array}}{6} \quad -\dfrac{\begin{array}{c}6\\2\end{array}}{4} \quad -\dfrac{\begin{array}{c}6\\4\end{array}}{2} \end{array}$$

2.

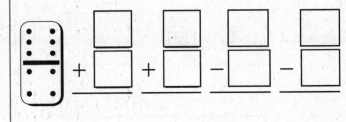

3.

4.

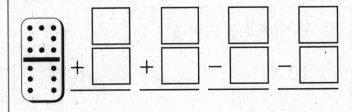

5.

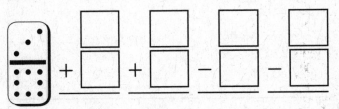

6.

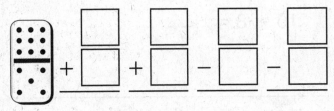

Try This

7.

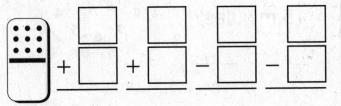

8. Write one addition fact and one subtraction fact.

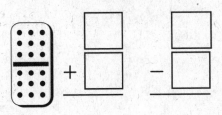

Math Boxes

1. Fill in the missing numbers.

144		
	155	

2. What is the temperature? Fill in the circle next to the best answer.

Ⓐ 55 degrees

Ⓑ 62 degrees

Ⓒ 52 degrees

Ⓓ 56 degrees

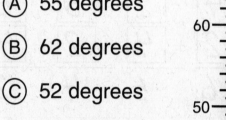

3. Write the sums.

$10 + 5 =$ _____

$10 + 6 =$ _____

$10 + 7 =$ _____

$10 + 8 =$ _____

Unit

4. Write these numbers in order from smallest to largest. Begin with the smallest number.

133, 146, 129, 151

_____ , _____ , _____ , _____

5. Put an X on the digit in the tens place.

456

309

6. What time is it?

_____ : _____

What time will it be in 15 minutes?

_____ : _____

LESSON 2·7

Subtraction Number Stories

Solve each problem.

1. Dajon has $11. He buys a book for $6. How much money does he have left?

   $ _____

2. Martin has 7 markers. Carlos has 4 markers. How many more markers does Martin have than Carlos?

   _____ markers

3. There are 11 girls on Tina's softball team. There are 13 girls on Lisa's team. How many more girls are on Lisa's team than on Tina's?

   _____ girls

4. Julia has 10 flowers. She gives 4 flowers to her sister. How many flowers does she have left?

   _____ flowers

5. Keisha has 8 chocolate cookies and 5 vanilla cookies. How many more chocolate cookies does she have than vanilla cookies?

   _____ chocolate cookies

6. Make up and solve your own subtraction story.

   _____

   _____

LESSON
2·7
Math Boxes

---

1. Use your Pattern-Block Template. Draw a rhombus.

There are _____ sides.

MRB
55

2. Circle the activity that takes about 1 minute.

Brushing your teeth.

Eating lunch.

Playing a soccer game.

---

3. Write the fact family for this domino.

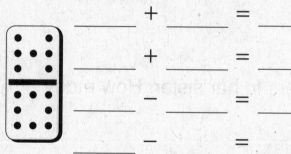

____ + ____ = ____

____ + ____ = ____

____ − ____ = ____

____ − ____ = ____

MRB
25

4. How likely is it that the school will serve lunch today? Circle your answer.

impossible

unlikely

likely

certain

---

5. Arrange the numbers in order from smallest to largest. Circle the middle number.

98, 56, 143, 172, 81

____, ____, ____, ____, ____

MRB
46

6. What time is it?

____ : ____

What time will it be in 30 minutes?

____ : ____

MRB
80 81

LESSON 2·8   Using a Pan Balance and a Spring Scale

Weighing Things with a Pan Balance

1. Pick two objects. Which feels heavier?

   _____

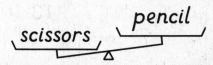

2. Put one of these objects in the left pan of the pan balance.

3. Put the other object in the right pan.

4. Show what happened on one of the pan-balance pictures.

   ◆ Write the names of the objects on the pan-balance picture.

   ◆ Draw a circle around the pan with the heavier object.

5. Repeat with other pairs of objects.

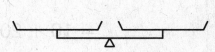

Weighing Things with a Spring Scale

1. Which is heavier: 1 ounce or 1 pound? _____

2. How many ounces are in 1 pound? _____

3. Put objects in the plastic bag on the spring scale.

4. Weigh them. Try to get a total weight of about 1 pound.

5. List the objects in the bag that weigh a total of about 1 pound.

_____   _____

_____   _____

Math Boxes

1. Fill in the missing numbers.

	13	

2. What temperature is it? Fill in the circle next to the best answer.

°F 70

Ⓐ 61

Ⓑ 62

Ⓒ 64

Ⓓ 78

60

50

3. Write the sums.

Unit _____

$10 + 7 =$ _____

$10 +$ _____ $= 12$

_____ $= 10 + 20$

_____ $= 10 + 41$

4. Write these numbers in order from smallest to largest. Circle the smallest number and draw a box around the largest number.

243, 156, 326, 256

_____, _____, _____, _____

5. Put an X on the digit in the tens place in each number.

362   1,043

1,209   596

6. What time is it?

_____ : _____

What time will it be in 15 minutes?

_____ : _____

LESSON 2·9   Name-Collection Boxes

1. Write 10 names in the 12 box.

12

2. Circle the names that DO NOT belong in the 9 box.

9

$12 - 3$ $8 + 0$

$9 - 0$ $5 + 4 + 1$

$19 - 10$ $\cancel{||||} \ |||$

$15 - 7$ x x x
   x x x
   x x x | 1 less than 10 |

$3 + 3 + 3$ nine

3. Three names DO NOT belong in this box. Circle them. Write the name of the box on the tag.

$9 + 3$ $12 - 8$

$3 + 3$ $\cancel{||||} \ ||$

x x x
x x x $5 + 3 - 2$

$10 - 4$ | half a dozen |

4. Make up a name-collection box of your own.

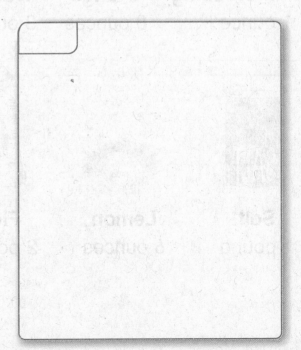

LESSON 2·9 Pan-Balance Problems

Reminder: There are 16 ounces in 1 pound.

Some food items and their weights are shown below.

◆ Pretend you will put one or more items in each pan.

◆ Pick items that would make the balances tilt the way they are shown on journal page 43.

◆ Write the name of each item in the pan you put it in.

◆ Write the weight of each item below the pan you put it in.

Try to use a variety of food items.

| **Salad Dressing** | **Orange** | **Walnuts** | **Eggplant** | **Gummy Worms** |
| 1 ounce | 8 ounces | 3 ounces | 15 ounces | 4 ounces |

| **Salt** | **Lemon** | **Flour** | **Banana** | **Potatoes** |
| 1 pound | 6 ounces | 2 pounds | 6 ounces | 5 pounds |

Pan-Balance Problems *continued*   43

Example:

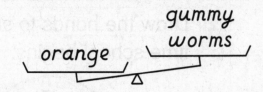

Weight: __8 ounces__   Weight: __4 ounces__

1. Weight: _____   Weight: _____

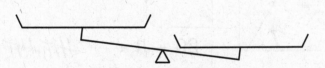

2. Weight: _____   Weight: _____

3. Weight: _____   Weight: _____

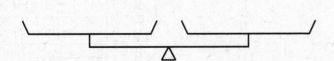

4. Weight: _____   Weight: _____

LESSON 2·9 Math Boxes

1. How much money?

$ _____ . _____

2. Draw the hands to show the time school begins.

3. Write the fact family.

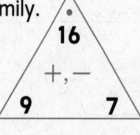

16

+, −

9   7

_____ + _____ = _____

_____ + _____ = _____

_____ − _____ = _____

_____ − _____ = _____

MRB 26 27

4. Write the label and add 3 more names.

20 − 4   ‖‖‖‖ ‖‖‖‖ ‖‖‖‖ |

MRB 16

5. Use your Pattern-Block Template to draw a trapezoid.

MRB 55

6. How many books in all did Pedro read on Saturday and Sunday? Fill in the circle next to the best answer.

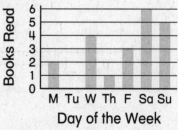

Ⓐ 10   Ⓑ 5   Ⓒ 6   Ⓓ 11

LESSON 2·10   **Frames-and-Arrows Problems**

1. Fill in the empty frames.

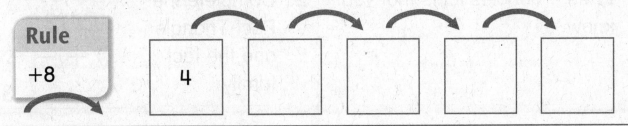

Rule: +8 | 4 | ☐ | ☐ | ☐ | ☐

2. Fill in the empty frames.

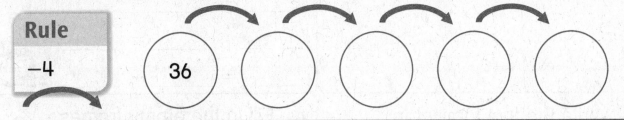

Rule: −4 | 36 | ○ | ○ | ○ | ○

3. Fill in the empty frames.

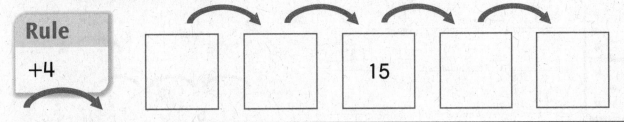

Rule: +4 | ☐ | ☐ | 15 | ☐ | ☐

4. Fill in the arrow rule.

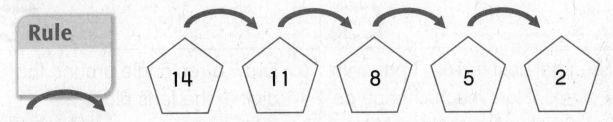

Rule: ___ | 14 | 11 | 8 | 5 | 2

Try This

5. Fill in the arrow rule and the empty frames.

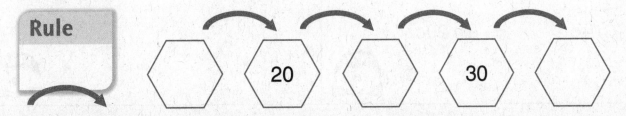

Rule: ___ | ⬡ | 20 | ⬡ | 30 | ⬡

LESSON
2·10 **Math Boxes**

---

1. Write 4 doubles facts that you know.

---

2. Complete the Fact Triangle and the fact family.

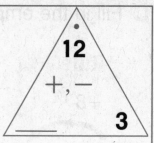

_____ = _____ + _____

_____ + _____ = _____

_____ − _____ = _____

_____ = _____ − _____

MRB
26 27

---

3. Write the fact family for the domino.

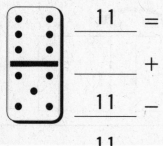

11 = _____ + _____

_____ + _____ = 11

11 − _____ = _____

11 − _____ = _____

MRB
25

---

4. Fill in the empty frames.

Rule
+2

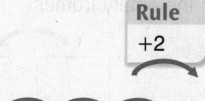

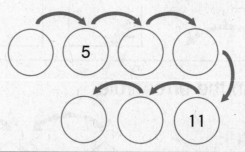

MRB
98

---

5. The total cost is 16¢. I pay with 2 dimes. How much change do I get? Fill in the circle next to the best answer.

Ⓐ 36¢   Ⓑ 6¢

Ⓒ 4¢   Ⓓ 20¢

MRB
88

---

6. Draw a rectangle around the digit in the tens place.

3 4 9

4 0 6

MRB
10

LESSON 2·11   **"What's My Rule?"**

In Problems 1–4, follow the rule. Fill in the missing numbers.

1.  Rule +6

in	out
2	8
3	9
5	
9	

2. 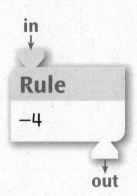 Rule −4

in	out
6	2
8	
10	
	5

3. 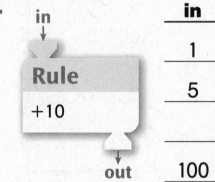 Rule +10

in	out
1	
5	15
	20
100	

4.  Rule −5

in	out
6	
	3
5	
12	

What is the rule? Write it in the box. Then fill in any missing numbers.

5.  Rule

in	out
6	13
1	8
3	
4	

6.  Rule

in	out
12	10
6	4
11	
	6

LESSON 2·11

Math Boxes

---

1. Show $1.50 three ways. Use Ⓠ, Ⓓ, and Ⓝ.

2. What time is it?

_____ : _____

What time will it be in 20 minutes?

_____ : _____

---

3. Write the fact family.

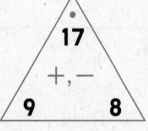

17

+,−

9   8

_____ + _____ = _____

_____ + _____ = _____

_____ − _____ = _____

_____ − _____ = _____

4. Write 6 names for 15.

15

---

5. Draw a hexagon with your Pattern-Block Template.

There are _____ sides.

6. How many cups of lemonade did Dee sell in the third hour? _____

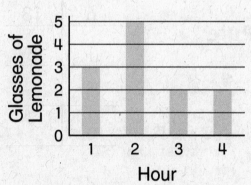

LESSON 2·12

Math Boxes

1. Write the doubles fact.

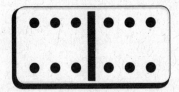

Number model:

_____ + _____ = _____

2. Fill in the sum on the Fact Triangle. Write the fact family.

8   5

_____ + _____ = _____

_____ + _____ = _____

_____ − _____ = _____

_____ − _____ = _____

3. Complete the fact family.

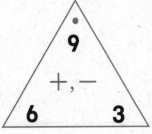

9

+,−

6   3

_____ = _____ + _____

_____ = _____ + _____

_____ = _____ − _____

_____ = _____ − _____

MRB 26 27

4. Fill in the frames.

Rule
+7

14   ◯

35

MRB 98

5. The total cost is 75¢. I pay with $1.00. How much change do I get?

_____

6. Circle the digit in the hundreds place.

128   972

465   2,425

MRB 10

LESSON 2·13 **Subtract 9 or 8**

> **Reminder:** To find $18 - 9$, think $18 - 10 + 1$.
>
> To find $18 - 8$, think $18 - 10 + 2$.

1. Subtract. Use the −9 and −8 shortcuts.

a. $13 - 9 =$ _____   b. $16 - 9 =$ _____   c. $14 - 8 =$ _____

d. _____ $= 12 - 8$ e. _____ $= 17 - 9$ f. $12 - 9 =$ _____

g. _____ $= 13 - 8$ h. $11 - 9 =$ _____   i. _____ $= 15 - 8$

j. $\begin{array}{r} 15 \\ -\ 9 \\ \hline \end{array}$ k. $\begin{array}{r} 17 \\ -\ 8 \\ \hline \end{array}$ l. $\begin{array}{r} 11 \\ -\ 8 \\ \hline \end{array}$

Try This

2. Find the differences.

a. $43 - 9 =$ _____   b. $56 - 8 =$ _____   c. $65 - 9 =$ _____

d. $37 - 8 =$ _____   e. $45 - 9 =$ _____   f. $53 - 8 =$ _____

---

3. Solve.

a. $7 =$ _____ $- 9$ b. $6 =$ _____ $- 8$

LESSON 2·13   Math Boxes

1. How much money?

$_____._____

2. Show 8:50 P.M.

3. Find the turn-around facts.

3 + 4 = _____

4 + _____ = 7

8 + 5 = _____

5 + _____ = 13

4. Write the label and add 3 more names.

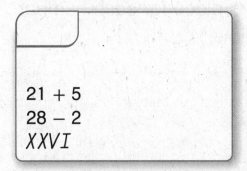

21 + 5
28 − 2
XXVI

16

5. Use your Pattern-Block Template to draw a trapezoid.

Circle the three sides that are the same length.

MRB
55

6. What day did Molly swim the most laps? Fill in the circle next to the best answer.

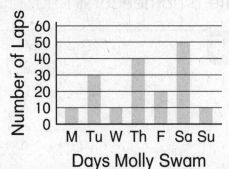

Number of Laps

Days Molly Swam

Ⓐ Wednesday   Ⓑ Sunday

Ⓒ Saturday   Ⓓ Tuesday

LESSON 2·14   **Math Boxes**

---

1. Selling Tickets for the School Fair

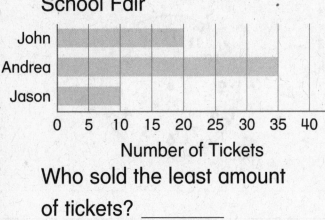

Number of Tickets

Who sold the least amount of tickets? _____

---

2. Use the Pattern-Block Template to draw a hexagon.

54

---

3. Put an X on the digit in the tens place for each number below.

95

145

217

1,273

10

---

4. Draw the hands to show 7:45 A.M.

---

5. Write 6 names for $1.00.

$1.00

---

6. Fill in the missing frames.

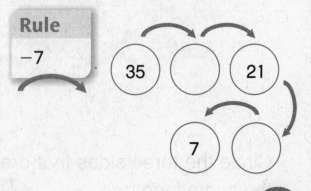

Rule

−7

35   21

7

98 99

LESSON 3·1 Place Value

Write the number for each group of base-10 blocks.

1. 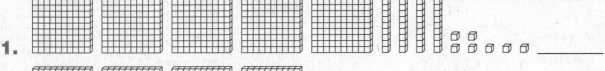 _____

2.  _____

3. Write a number with …

5 in the ones place,

3 in the hundreds place, and

2 in the tens place. _____

4. 506

How many hundreds? _____

How many tens? _____

How many ones? _____

Write <, >, or =.

5. 328 _____ 322   **6.** 122 _____ 102   **7.** 623 _____ 633

8. Marta wrote 24 to describe the number shown by these base-10 blocks:

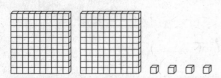

Is Marta right? Explain your answer.

_____

_____

_____

LESSON 3·1   **Math Boxes**

1. Fill in the frames. Use your calculator to count by 7s.

Rule
+7

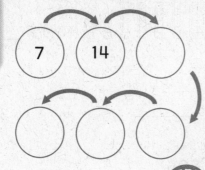

7   14

MRB
98 99

2. Write <, >, or =.

9 + 7 _____ 13

10 + 12 _____ 26

7 + 7 _____ 5 + 9

MRB
9

3. Solve.

Unit

7 + 8 = _____

70 + 80 = _____

700 + 800 = _____

7,000 + 8,000 = _____

4. How much money? Fill in the circle next to the best answer.

Ⓐ $21.40   Ⓑ $21.45

Ⓒ $2.45   Ⓓ $11.40

5. Mike had 7¢. He found a dime. How much does he have now?

Change

Start		End

Answer: _____ ¢

MRB
116

6. Solve.

Unit

18 + 9 = _____

17 + 9 = _____

16 + 9 = _____

15 + 9 = _____

LESSON 3·2   *Spinning for Money*

Materials   ☐ *Spinning for Money* Spinner (*Math Masters*, p. 472)

☐ pencil   ☐ large paper clip

☐ 7 pennies, 5 nickels, 5 dimes, 4 quarters, and one $1 bill for each player

☐ sheet of paper labeled "Bank"

Players   2, 3, or 4

Skill   Exchange coins and dollar bills

Object of the Game To be first to exchange for a $1 bill

Directions

1. Each player puts 7 pennies, 5 nickels, 5 dimes, 4 quarters, and one $1 bill into the bank.

2. Players take turns spinning the *Spinning for Money* Spinner and taking the coins shown by the spinner from the bank.

3. Whenever possible, players exchange coins for a single coin or bill of the same value. For example, a player could exchange 5 pennies for a nickel, or 2 dimes and 1 nickel for a quarter.

4. The first player to exchange for a $1 bill wins.

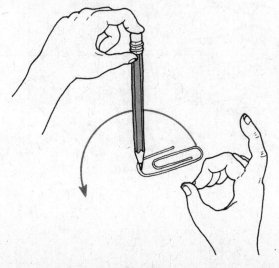

Use a large paper clip and pencil to make a spinner.

LESSON 3·2 Fruit and Vegetables Stand Poster

LESSON 3·2 Buying Fruit and Vegetables

Select the fruit and vegetables from journal page 56
that you would like to buy. Write the name of each item.

Then draw the coins you could use to pay for each item.
Write Ⓟ, Ⓝ, Ⓓ, or Ⓠ.

For Problems 3 and 4, write the total amount of money
that you would spend.

I bought (Write the name.)	I paid (Draw coins.)	I paid (Draw coins another way.)
Example: one <u>*orange*</u>	Ⓓ Ⓝ Ⓟ Ⓟ Ⓟ	Ⓝ Ⓝ Ⓟ Ⓟ Ⓟ Ⓟ Ⓟ Ⓟ Ⓟ Ⓟ
1. one _____		
2. one _____		
3. one _____ and one _____		Total: _____
Try This		
4. one _____, one _____, and one _____		Total: _____

LESSON 3·2 Math Boxes

1. Draw hands to show 4:30.

2. Write seven even numbers.

_____ _____ _____

_____ _____ _____

_____

MRB 97

3. Put these numbers in order from smallest to largest and circle the middle number.

23, 59, 49, 3, 159

_____, _____, _____, _____, _____

4. Fill in the tally chart. Grade 1 sold 17 cupcakes at the bake sale and Grade 2 sold 13 cupcakes.

Number of Cupcakes Sold	
Grade 1	
Grade 2	

MRB 40

5. What is the temperature?

_____

Is it warm or cold?

_____

MRB 87

6. A bag of pretzels costs 95¢. About how much money would you need to buy 3 bags of pretzels? Fill in the circle next to the best answer.

Ⓐ $1.50   Ⓑ 95¢

Ⓒ $3.00   Ⓓ $3.95

What Time Is It?

1. Write the time.

____ : ____   ____ : ____   ____ : ____   ____ : ____

2. Draw the hands to match the time.

8:00   3:30   7:45   9:15

3. Make up times of your own. Draw the hands to show each time. Write the time under each clock.

____ : ____   ____ : ____   ____ : ____

LESSON 3·3 **Math Boxes**

---

1. Use your calculator to count by 6s. Fill in the frames.

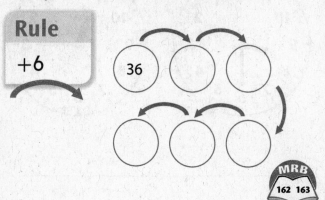

Rule
+6

36

MRB
162 163

---

2. Write <, >, or =.

6 + 7 _____ 15 − 4

5 + 8 _____ 8 + 5

18 − 9 _____ 5 + 4

MRB
9

---

3. Solve.

Unit

2 + 5 = _____

20 + 50 = _____

200 + 500 = _____

2,000 + 5,000 = _____

---

4. How much?

  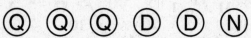

_____

---

5. Chen had 30 postcards. He collected 17 more postcards. How many does he have now?

Change

Start _____ End

Answer: _____ postcards

---

6. Write the number that is 10 more.

19 _____

29 _____

39 _____

49 _____

LESSON 3·4 **Build a Number**   **61**

Your number	Show your number using base-10 blocks	Show your number another way using base-10 blocks

LESSON 3·4 Geoboard Dot Paper (7 × 7)

1.

2.

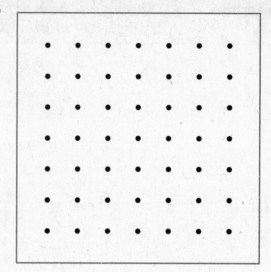

3.

4.

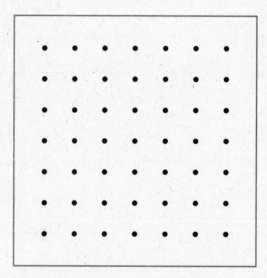

5.

6.

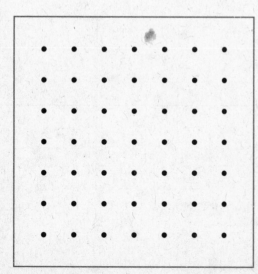

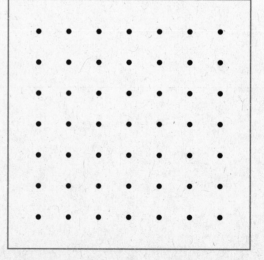

LESSON 3·4   **Magic Squares**   63

1. Add the numbers in each row. Add the numbers in each column.
    Add the numbers on each diagonal.

    Are the sums all the same? _____

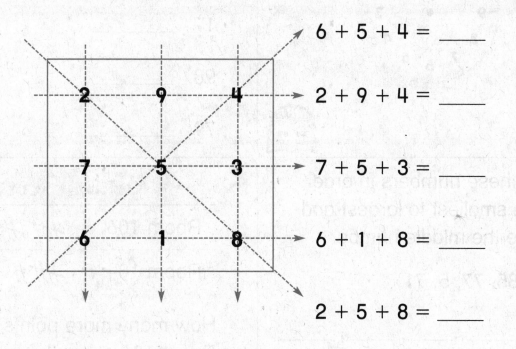

$6 + 5 + 4 =$ ____

$2 + 9 + 4 =$ ____

$7 + 5 + 3 =$ ____

$6 + 1 + 8 =$ ____

$2 + 5 + 8 =$ ____

____ ____ ____

2. The sum of each row, column, and diagonal must be 15.
    Find the missing numbers. Write them in the blank boxes.

	7	
9		1
4		8

8		6
3		

LESSON 3·4 **Math Boxes**

---

1. Draw hands to show 7:30.

MRB 80 81

---

2. Write odd or even next to each number.

5 _____

17 _____

98 _____

79 _____

MRB 97

---

3. Put these numbers in order from smallest to largest and circle the middle number.

12, 85, 77, 5, 71

____ , ____ , ____ , ____ , ____

MRB 46

4.

Kickball Score	
Room 106	
Room 104	

How many more points did Room 106 score than Room 104?

_____

---

5. What is the temperature?

_____

Is it warm or cold?

_____

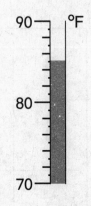

MRB 87

---

6. A slice of pizza costs $2.12. A candy bar costs 52¢. If Jaleesa has $3.00, does she have enough money to buy the candy bar and the pizza?

_____

---

64 sixty-four

LESSON 3·5 Graphing Pockets Data

Count the pockets of children in your class.

Pockets	Children	
	Tallies	Number
0		
1		
2		
3		
4		
5		
6		
7		
8		
9		
10		
11		
12		
13 or more		

Graphing Pockets Data

Draw a picture graph of the pockets data.

KEY: Each 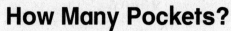 = 1 child

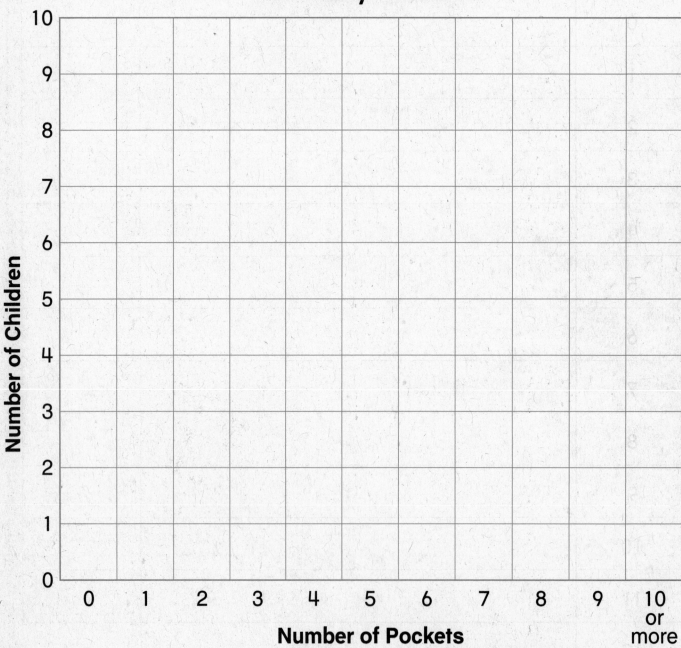

LESSON
3·5

Graphing Pockets Data

Draw a bar graph of the pockets data.

How Many Pockets?

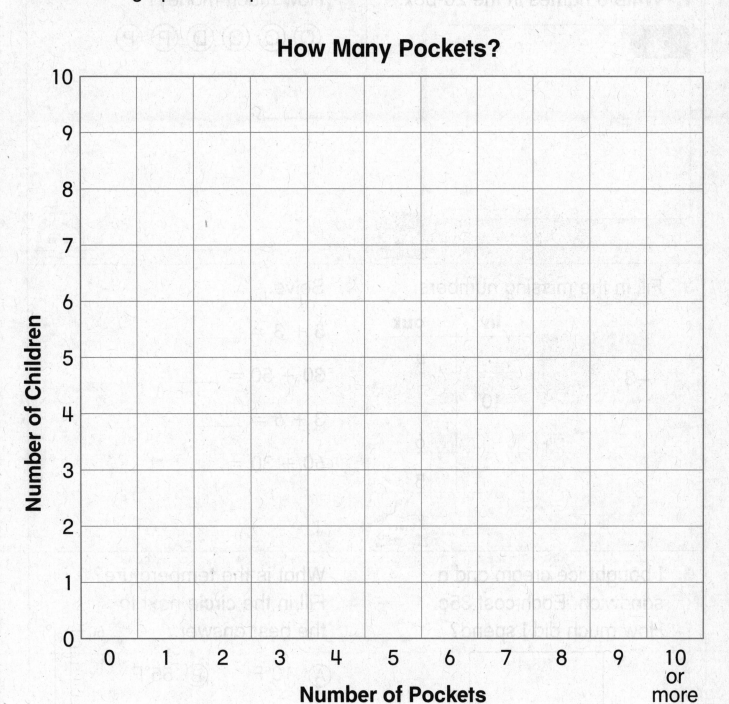

Number of Children

Number of Pockets

LESSON 3·5

Math Boxes

1. Write 6 names in the 20-box.

20

MRB 16

2. How much money?

Q Q Q D P P

_____ ¢

MRB 88 89

3. Fill in the missing numbers.

Rule
−3

in	out
	4
10	
	6
	5

MRB 101

4. Solve.

$5 + 3 =$ _____

$30 + 50 =$ _____

$3 + 6 =$ _____

$60 + 30 =$ _____

Unit

5. I bought ice cream and a sandwich. Each cost 35¢. How much did I spend?

Total	
Part	Part

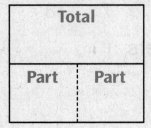

Answer: _____ ¢

6. What is the temperature? Fill in the circle next to the best answer.

Ⓐ 10°F   Ⓑ 55°F

Ⓒ 40°F   Ⓓ 50°F

60 — °F

50 —

40 —

LESSON 3·6   Two-Rule Frames and Arrows

Fill in the frames. Use coins to help you.

1.

Rule	
Add 5¢	

Rule	
Add 10¢	

5¢   10¢   20¢   25¢

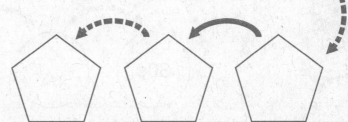

2.

Rule	
Add 10¢	

Rule	
Subtract 5¢	

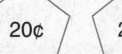

20¢   15¢

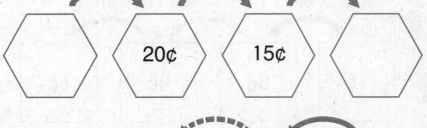

Try This

3.

Rule	
Add 25¢	

Rule	
Add 5¢	

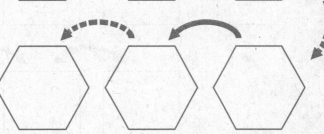

25¢   50¢   80¢

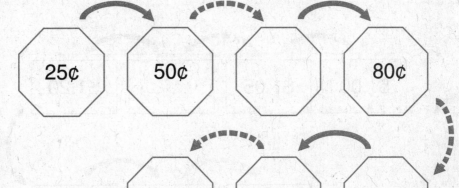

LESSON 3·6

Two-Rule Frames and Arrows *continued*

4. Fill in the frames. Use coins to help you.

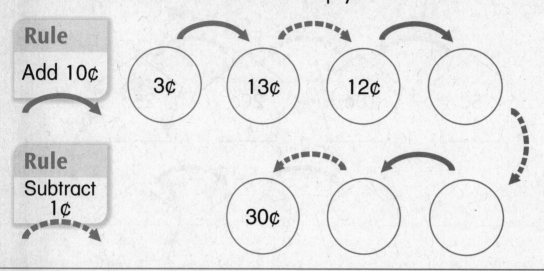

Fill in the frames and find the missing rules. Use coins to help you.

5.

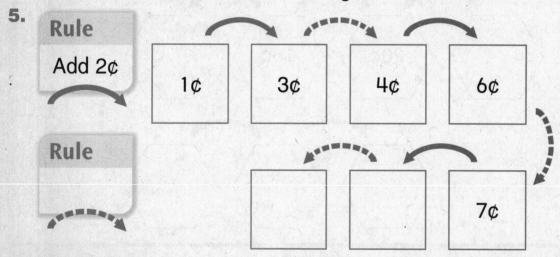

Try This

6.

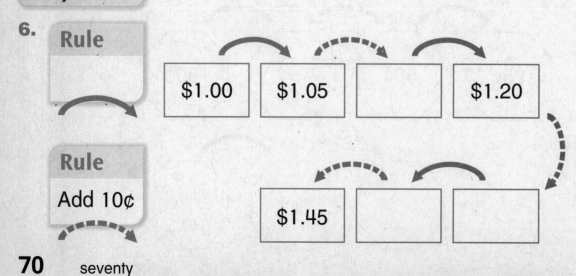

70 seventy

LESSON 3·6

Dollar Rummy

71

Materials
- ☐ *Dollar Rummy* cards (*Math Masters*, p. 454)
- ☐ scissors to cut out cards
- ☐ cards from *Math Masters*, p. 455 (optional)

Players   2

Skill   Find complements of 100

Object of the Game   To have more $1.00 pairs

Directions

1. Deal 2 *Dollar Rummy* cards to each player.

2. Put the rest of the deck facedown between the players.

3. Take turns. When it's your turn, take the top card from the deck. Lay it faceup on the table.

4. Look for two cards that add up to $1.00. Use any cards that are in your hand or faceup on the table.

5. If you find two cards that add up to $1.00, lay these two cards facedown in front of you.

6. When you can't find any more cards that add up to $1.00, it is the other player's turn.

7. The game ends when all of the cards have been used or when neither player can make a $1.00 pair.

8. The winner is the player with more $1.00 pairs.

LESSON 3·6

Math Boxes

1. Write the amount. Fill in the circle next to the best answer.

_____¢

Ⓐ 56¢

Ⓑ 55¢

Ⓒ 30¢

Ⓓ 46¢

MRB 88 89

2. Solve.

Unit

$70 + \_\_\_\_\_ = 100$

$100 = 20 + 30 + \_\_\_\_\_$

$\_\_\_\_\_ = 50 + 40 + 10$

$100 = \_\_\_\_\_ + 10 + 60$

3. How many children have cats?

_____

How many children have fish?

_____

Children's Pets

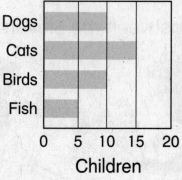

0   5   10   15   20
Children

4. Tomorrow I will get dressed for the day. (Circle one.)

Likely

Unlikely

Impossible

5. I had 17 tulips. I planted 20 more. How many do I have now?

Change

Start _____ End

Answer: _____ tulips

6. Solve.

Unit

$\_\_\_\_\_ = 70 - 60$

$\_\_\_\_\_ = 88 - 80$

$63 - 20 = \_\_\_\_\_$

$\_\_\_\_\_ = 92 - 30$

LESSON 3·7 Shopping at the Fruit and Vegetables Stand

Price per Item

pear	13¢	melon slice	30¢	lettuce	45¢
orange	18¢	apple	12¢	green pepper	24¢
banana	9¢	tomato	20¢	corn	15¢
plum	6¢	onion	7¢	cabbage	40¢

Complete the table.

I bought	I paid (Draw coins or $1 bill.)	I got in change
_____		_____ ¢
_____		_____ ¢
_____		_____ ¢

Try This

Buy 2 items. How much change from $1.00 will you get?

I bought	I paid	I got in change
_____ and _____	$1	_____ ¢

LESSON 3·7 **Math Boxes**

1. Write 3 names for 50.

> 50

2. Write the fewest number of coins needed to make 67¢.

67¢ = _____ quarters

_____ dimes

_____ nickels

_____ pennies

3. Fill in the rule and the missing numbers.

Rule

in	out
132	122
103	93
114	
205	

4. Solve.

Unit

6 + 8 = _____

80 + 60 = _____

7 + 4 = _____

40 + 70 = _____

5. 10 children ordered juice. 13 children ordered milk. How many children ordered drinks?

Total	
Part	**Part**

Answer: _____ children

6. What is the temperature?

_____

Would you wear a coat?

_____

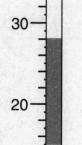

Making Change

LESSON 3·8

Materials ☐ 2 nickels, 2 dimes, 2 quarters, and one $1 bill
for each child

☐ 2 six-sided dice

☐ a cup, a small box, or a piece of paper for a bank

Number of children   2 or 3

Directions

1. Each person starts with 2 nickels, 2 dimes, 2 quarters, and one $1 bill.
   Take turns rolling the dice and finding the total number of dots that are
   faceup.

2. Use the chart to find out how much money to put in the bank. (There is
   no money in the bank at the beginning of the activity.)

Making Change Chart

Total for Dice Roll	2	3	4	5	6	7	8	9	10	11	12
Amount to Pay the Bank	10¢	15¢	20¢	25¢	30¢	35¢	40¢	45¢	50¢	55¢	60¢

3. Use your coins to pay the amount to the bank. You can
   get change from the bank.

4. Continue until someone doesn't have enough money
   left to pay the bank.

LESSON 3·8

Buying from a Vending Machine

1. The exact change light is on. You want to buy a carton of orange juice. Which coins will you put in? Draw Ⓝ, Ⓓ, and Ⓠ to show the coins.

2. The exact change light is off. You want to buy a carton of 2% milk. You don't have the exact change. Which coins or bills will you put in? Draw coins or a $1 bill.

How much change will you get? _____

Buying from a Vending Machine *continued*

3. The exact change light is on.

You buy:	Draw the coins you put in.
chocolate milk	
strawberry yogurt drink	
_____	

4. The exact change light is off.

You buy:	Draw the coins or the $1 bill you put in.	What is your change?
orange juice	Q Q Q	_____ ¢
chocolate milk	$1	_____ ¢
_____		_____ ¢
_____		_____ ¢
_____		_____ ¢

LESSON 3·8   Math Boxes

1. How much money?

    Ⓓ Ⓝ

Ⓓ Ⓝ

_____ ¢ or $ _____

88–90

2. Solve.

$100 = 75 +$ _____

$60 +$ _____ $= 100$

$$\begin{array}{r} 95 \\ +\ 5 \\ \hline \end{array} \qquad \begin{array}{r} 90 \\ +\ 10 \\ \hline \end{array}$$

Unit

3. How many children ate 2 scoops of ice cream?

_____

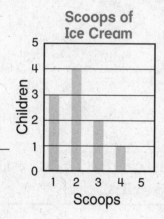

Scoops of Ice Cream

Children (y-axis 0–5)

Scoops (x-axis 1–5)

4. Tomorrow I will eat lunch. (Circle one.)

Likely

Unlikely

Impossible

5. How old will you be in 18 years?

Change

Start		End

Answer: _____ years old

6. Solve.

$23 + 30 =$ _____

$86 =$ _____ $+ 50$

_____ $= 40 + 59$

$67 + 30 =$ _____

Unit

LESSON 3·9 **Math Boxes**

---

1. A.M. temperature was 50°F.
P.M. temperature was 68°F.

What was the change? Fill in the diagram.

Change

Start		End
	_____	

Answer: _____°F

---

2. Mel picked 12 apples. Meg picked 18 apples. How many in all?

Total	
Part	**Part**
12	**18**

Answer: _____ apples

---

3. I have $1.00. Can I buy two 45¢ ice cream bars?

_____

---

4. What is the temperature?

_____

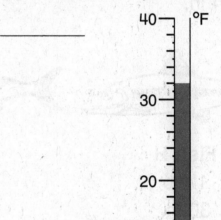

---

5. Solve.

$22 + 10 =$ _____

$17 + 12 =$ _____

Unit _____

$$\begin{array}{r} 45 \\ + 35 \\ \hline \end{array} \qquad \begin{array}{r} 100 \\ + 120 \\ \hline \end{array}$$

---

6. Solve.

$12 + 10 =$ _____

$13 + 10 =$ _____

Unit _____

$$\begin{array}{r} 29 \\ + 10 \\ \hline \end{array} \qquad \begin{array}{r} 34 \\ + 10 \\ \hline \end{array}$$

LESSON 4·1 **Fish Poster**

Fish A
1 lb
12 in.

Fish B
3 lb
14 in.

Fish C
4 lb
18 in.

Fish D
5 lb
24 in.

Fish E
6 lb
24 in.

Fish F
8 lb
30 in.

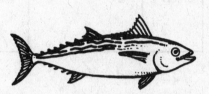

Fish G
10 lb
30 in.

Fish H
14 lb
30 in.

Fish I
15 lb
30 in.

Fish J
24 lb
36 in.

Fish K
35 lb
42 in.

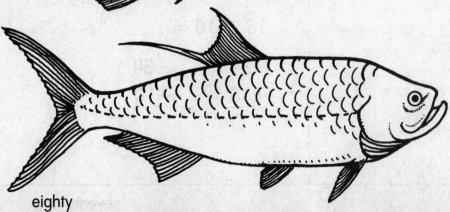

Fish L
100 lb
72 in.

LESSON 4·1 "Fishy" Stories

Use the information on journal page 80 for Problems 1–4.
Do the following for each number story:

◆ Write the numbers you know in the change diagram.

◆ Write "?" for the number you need to find.

◆ Answer the question.

◆ Write a number model.

---

1. Fish J swallows Fish B.

How much does Fish J weigh now?

Answer: _____ pounds

Number model: _____

Start [ ] → Change → End [ ]
_____

---

2. Fish K swallows Fish G.

How much does Fish K weigh now?

Answer: _____ pounds

Number model: _____

Start [ ] → Change → End [ ]
_____

---

3. Fish L swallows Fish F.

How much does Fish L weigh now?

Answer: _____ pounds

Number model: _____

Start [ ] → Change → End [ ]
_____

"Fishy" Stories *continued*

LESSON 4·1

Try This

4. Fish I swallows another fish.

   Fish I now weighs 20 pounds.

   Which fish did Fish I swallow?

   Answer: _____

   Number model: _____

5. Fish F swallows Fish B.

   Then Fish K swallows Fish F.

   How much does Fish K weigh now?

   Answer: _____ pounds

6. Write a number story that involves three different fish.

   _____

   _____

   _____

   _____

   _____

   Answer: _____ pounds

LESSON 4·1   # Distances on a Number Grid

Use the number grid below. Find the distance from the first number to the second. Start at the first number and count the number of spaces moved to reach the second number.

1. 53 and 58 _____

2. 64 and 56 _____

3. 69 and 99 _____

4. 83 and 63 _____

5. 77 and 92 _____

6. 93 and 71 _____

7. 84 and 104 _____

8. 106 and 88 _____

9. 94 and 99 _____

10. 85 and 76 _____

11. 58 and 108 _____

12. 107 and 57 _____

13. 61 and 78 _____

14. 72 and 53 _____

15. 52 and 100 _____

16. 100 and 78 _____

51	52	53	54	55	56	57	58	59	60
61	62	63	64	65	66	67	68	69	70
71	72	73	74	75	76	77	78	79	80
81	82	83	84	85	86	87	88	89	90
91	92	93	94	95	96	97	98	99	100
101	102	103	104	105	106	107	108	109	110

Math Boxes

1. Write the fact family.

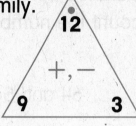

12

+, −

9   3

____ + ____ = ____

____ + ____ = ____

____ − ____ = ____

____ − ____ = ____

MRB
27

2. What time is it? Circle the best answer.

A. 6:30

B. 6:25

C. 6:27

D. 5:30

3. Write the number that is 100 more.

104 _____

204 _____

304 _____

404 _____

4. How much money?

| $20 | $20 |

| $10 | $1 |

Ⓠ Ⓠ Ⓠ Ⓓ

$ _____ . _____

5. Write <, >, or =.

2,000 _____ 1,000

2,500 _____ 3,500

4,321 _____ 1,234

5,674 _____ 5,674

MRB
9

6. A baseball is shaped like a sphere. Name something else that is shaped like a sphere.

_____

MRB
57

LESSON 4·2 Parts-and-Total Number Stories

Lucy's Snack Bar Menu					
Sandwiches		**Drinks**		**Desserts**	
Hamburger	65¢	Juice	45¢	Apple	15¢
Hot dog	45¢	Milk	35¢	Orange	25¢
Cheese	40¢	Soft drink	40¢	Banana	10¢
Peanut butter and jelly	35¢	Water	25¢	Cherry pie	40¢

For Problems 1–4, you are buying two items. Use the diagrams to record both the cost of each item and the total cost.

1. a soft drink and a banana

Total	
Part	**Part**

2. a hot dog and an apple

Total	
Part	**Part**

3. a soft drink and a slice of pie

Total	
Part	**Part**

4. a hamburger and juice

Total	
Part	**Part**

Try This

5. Jean buys milk and an orange. The cost is _____.

Jean gives the cashier 3 quarters.

How much change does she get? _____

LESSON 4·2   **Math Boxes**

---

1. Solve.

Unit

$8 + 7 =$ _____

$80 + 70 =$ _____

$800 + 700 =$ _____

$8,000 + 7,000 =$ _____

---

2. A piece of candy costs 11¢. I pay with 15¢. How much change do I get? Circle the best answer.

A. 26¢   **B.** 4¢

C. 5¢   **D.** 6¢

---

3. Estimate.

Is 7 closer to 0 or closer to 10? _____

Is 53 closer to 50 or closer to 60? _____

Is 88 closer to 80 or closer to 90? _____

---

4. Circle names that belong.

$1.00

10 dimes

4 quarters   18 nickels

100 pennies

5 dimes   5 nickels

---

5. Circle the number sentences that are true.

$9 + 7 = 7 + 9$

$8 - 5 = 5 - 8$

$6 + 5 = 5 + 6$

---

6. Draw the other half of the shape and write the name of it.

_____

LESSON 4·3 Temperatures

Fahrenheit Thermometer

°F

- 230
- 220
- 210
- 200
- 190
- 180
- 170
- 160
- 150
- 140
- 130
- 120
- 110
- 100
- 90
- 80
- 70
- 60
- 50
- 40
- 30
- 20
- 10
- 0
- −10

Water boils. 212°F, 100°C

Normal body temperature 98.6°F, 37°C

Room temperature 70°F, 21°C

Water freezes. 32°F, 0°C

Celsius Thermometer

°C

- 110
- 100
- 90
- 80
- 70
- 60
- 50
- 40
- 30
- 20
- 10
- 0
- −10
- −20

1. Use a thermometer to measure and record the temperatures of the following:

   a. your classroom _____°F _____°C

   b. hot water from a faucet _____°F _____°C

   c. ice water _____°F _____°C

2. Which temperature makes more sense? Circle it.

   a. temperature in a classroom:

      40°F or 70°F

   b. temperature of hot tea:

      100°F or 180°F

   c. temperature of a person with a fever:

      100°F or 100°C

   d. temperature on a good day for ice-skating outside:

      −10°C or 10°C

LESSON 4·3 **Math Boxes**

---

1. Write the fact family.

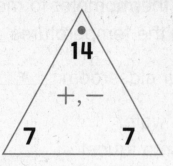

14

+, −

7   7

_____ + _____ = _____

_____ − _____ = _____

---

2. What time is it?

_____

What time will it be in 1 hour?

_____

---

3. Count forward by 100s.

25, _____, 225, _____,

_____, 525, 625

---

4. Show two ways to make 85¢. Use Ⓟ, Ⓝ, Ⓓ, and Ⓠ.

---

5. Write <, >, or =.

1,002 _____ 102

3,700 _____ 7,300

2,310 _____ 2,410

5,697 _____ 5,696

---

6. Your pencil tip is shaped like a cone. Name something else that is shaped like a cone.

_____

LESSON 4·4 Parts-and-Total Number Stories

For each number story:

◆ Write the numbers you know in the parts-and-total diagram.

◆ Write "?" for the number you want to find.

◆ Answer the question. Remember to include the unit.

◆ Write a number model.

1. Jack rode his bike for 20 minutes on Monday. He rode it for 30 minutes on Tuesday. How many minutes did he ride his bike in all?

   Answer: _____

   (unit)

   Number model: _____

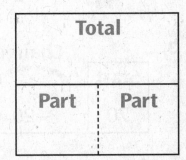

Total	
Part	**Part**

2. Two children collect stamps. One child has 40 stamps. The other child has 9 stamps. How many stamps do the two children have together?

   Answer: _____

   (unit)

   Number model: _____

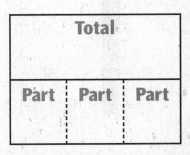

Total	
Part	**Part**

Try This

3. 25 children take ballet class. 15 children take art class. 5 children take a sports class. In all, how many children take the three classes?

   Answer: _____

   (unit)

   Number model: _____

Total		
Part	**Part**	**Part**

LESSON
4·4

Temperature Changes

Write the missing number in each End box.

Then fill in the End thermometer to show this number.

Unit
°F

1.

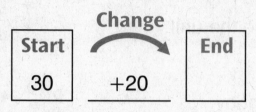

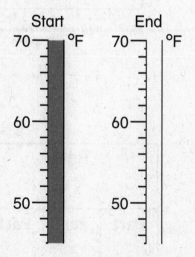

2.

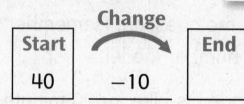

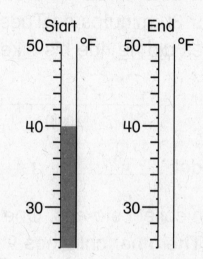

3.

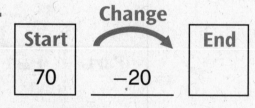

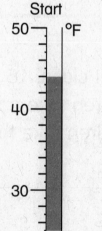

4.

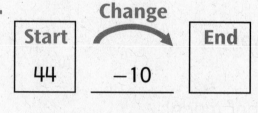

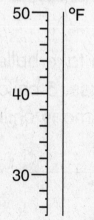

LESSON 4·4 **Temperature Changes** *continued*

Fill in the missing numbers for each diagram.

5.

6.

7.

Start _____ Change _____ End

8.

Start _____ Change _____ End

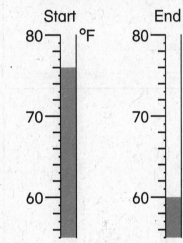

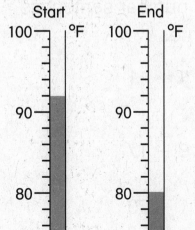

LESSON 4·4   Math Boxes

---

1. Solve.

Unit

_____ = 6 + 5

16 + 5 = _____

46 + 5 = _____

_____ = 76 + 5

86 + 5 = _____

---

2. LaVon has $1.00 and spends 73¢. How much change does she get?

_____

---

3. Estimate.

Is 49 closer to 40 or closer to 50? _____

Is 121 closer to 120 or closer to 130? _____

Is 214 closer to 210 or closer to 220? _____

---

4. Write 3 names for $2.00.

$2.00

---

5. Circle the number sentences that are true.

11 + 4 = 4 + 11

17 + 8 = 8 + 17

20 − 1 = 1 − 20

---

6. Draw the other half of the shape and write the name of it.

_____

LESSON 4·5 School Supply Store

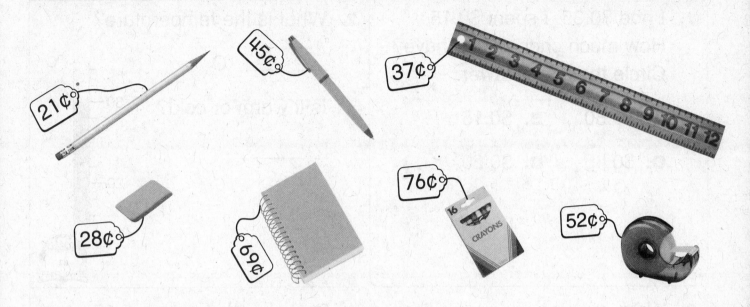

You have $1.00 to spend at the School Store.
Use estimation to answer each question.

Can you buy: **Write *yes* or *no*.**

1. a notebook and a pen?   _____

2. a pen and a pencil?   _____

3. a box of crayons and a roll of tape?   _____

4. a pencil and a box of crayons?   _____

5. 2 rolls of tape?   _____

6. a pencil and 2 erasers?   _____

7. You want to buy two of the same item.
List items you could buy two of with $1.00.

_____   _____

_____   _____

8. How many pencils could you buy with $1.00? _____

LESSON 4·5 Math Boxes

---

1. I had $0.35. I spent $0.15. How much change do I have? Circle the best answer.

A. $0.50   **B.** $0.15

C. $0.45   **D.** $0.20

---

2. What is the temperature?

_____°C

Is it warm or cold?

_____

---

3. Write *even* or *odd* on the line.

23 _____

52 _____

258 _____

197 _____

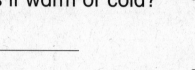

---

4. Draw hands to show 5:15.

---

5. Fill in the frames.

Rule
−2

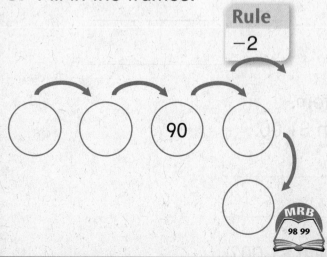

---

6. Use your Pattern-Block Template. Draw a trapezoid.

How many sides? ___ sides

LESSON 4·6 **Shopping Activity**

Do this activity with a partner.

Materials
- ☐ Shopping cards (*Math Masters,* p. 105)
- ☐ *Math Masters,* p. 433
- ☐ calculator
- ☐ play money for each partner: at least nine $1 bills and eight $10 bills; one $100 bill (optional)

Directions

1. Partners take turns being the Customer and the Clerk.

2. The Customer draws two cards and turns them over. These are the items the Customer is buying.

3. The Customer places the two cards on the parts-and-total diagram— one card in each Part box.

4. The Customer figures out the total cost of the two items without using a calculator.

5. The Customer counts out bills equal to the total cost. The Customer places the bills in the Total box on the parts-and-total diagram.

6. The Clerk uses a calculator to check that the Customer has figured the correct total cost.

7. Partners switch roles.

8. Play continues until all eight cards have been used.

Another Way to Do the Activity

Instead of counting out bills, the Customer says or writes the total. The Customer gives the Clerk a $100 bill to pay for the items. The Clerk must return the correct change.

LESSON 4·6 Shopping Poster

| Telephone $46 | Camera $43 | CD Player $25 | Calculator $17 |

| Toaster $29 | Iron $32 | CD $14 | Radio $38 |

LESSON 4·6 **Shopping Problems**

1. You buy a telephone and an iron. What is the total cost? $ _____

   Number Model: _____

Total	
Part	**Part**
46	32

2. You buy a radio and a calculator. What is the total cost? $ _____

   Number Model: _____

Total	
Part	**Part**
38	17

Solve each problem.

3. You bought two items. The total cost is exactly $60. What did you buy?

   _____

   _____

4. You bought two items. Together they cost the same as a telephone. What did you buy?

   _____

   _____

Try This

5. You bought three items. The total cost is exactly $60. What did you buy?

   _____

   _____

Math Boxes

1. 45 cents = 1 quarter
and _____ dimes

60 cents = 3 dimes
and _____ nickels

2. A.M. temperature was 50°F.
P.M. temperature is 68°F.

What was the change? _____°F
Fill in the diagram and
write the number model.

Change

Start		End

_____

_____

MRB
116–118

3. 20 airplanes. 8 take off.

How many stay? _____ airplanes
stay. Fill in the diagram and
write a number model.

Change

Start _____ End

_____

MRB
116–118

4. Draw hands to show 8:15.

5. Measure the length of this line
segment. Circle the best answer.

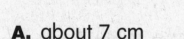

A. about 7 cm

B. about 6 cm

C. about 8 cm

D. about 10 cm

6. Draw a rectangle. Two sides
are 4 cm long and two sides
are 3 cm long.

MRB
54

LESSON 4·7   Measuring Lengths with a Tape Measure

1. Measure the height from the top of your desk or table to the floor. Measure to the nearest inch and centimeter.

   The height from my desk or table to the

   floor is about _____ inches.

   The height from my desk or table to the

   floor is about _____ centimeters.

2. Measure the height from the top of your chair to the floor. Measure to the nearest inch and centimeter.

   The height from the top of my chair to the

   floor is about _____ inches.

   The height from the top of my chair to the

   floor is about _____ centimeters.

3. Measure the width of your classroom door.

   The classroom door is about

   _____ inches wide.

   The classroom door is about

   _____ centimeters wide.

LESSON 4·7 Measuring Lengths with a Tape Measure *cont.*

4. Open your journal so it looks like the drawing below.

a. Measure the long side and the short side to the nearest inch.

The long side is about _____ inches.

The short side is about _____ inches.

b. Now measure your journal to the nearest centimeter.

The long side is about _____ centimeters.

The short side is about _____ centimeters.

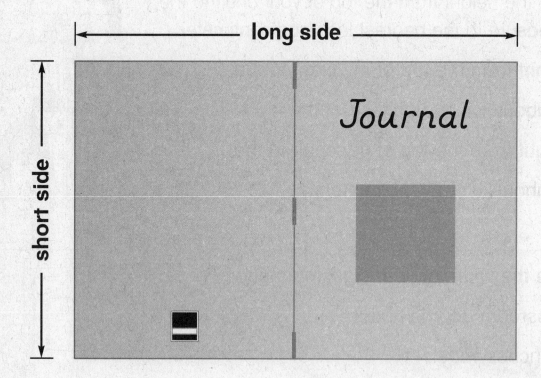

Tiling Surfaces with Shapes

Materials
- ☐ square pattern blocks
- ☐ slates
- ☐ sheets of paper
- ☐ Pattern-Block Template
- ☐ scissors
- ☐ Everything Math Deck cards, if available

1. Use square pattern blocks. **Tile** a card by covering it with the square pattern blocks.

   ◆ Lay the blocks flat on the card.

   ◆ Don't leave any spaces between blocks.

   ◆ Keep the blocks inside the edges of the card. There may be open spaces along the edges.

   Count the blocks on the card. If a space could be covered by more than half of a block, count the space as one block. Do not count spaces that could be covered by less than half of a block.

   Number of square pattern blocks needed to tile the card:

   _____

   Trace the card. Use your Pattern-Block Template to draw the square pattern blocks you used to tile the card.

LESSON 4·7 **Tiling Surfaces with Shapes** *continued*

2. Use Everything Math Deck cards to tile both a slate and a Pattern-Block Template. How many cards were needed to tile them?

Slate: _____ cards

Pattern-Block Template: _____ cards

3. Fold a sheet of paper into fourths. Cut the fourths apart. Use them to tile larger surfaces, such as a desktop.

Surface	Number of Fourths
_____	_____
_____	_____

Follow-Up

With a partner, find things in the classroom that are tiled or covered with patterns. Make a list. Be ready to share your findings.

LESSON 4·7 — An Attribute Rule

Choose an Attribute Rule Card. Copy the rule below.

Rule: _____

Draw or describe all the attribute blocks that fit the rule.

Draw or describe all the attribute blocks that do *not* fit the rule.

These blocks fit the rule:

_____

_____

_____

_____

These blocks do *not* fit the rule:

_____

_____

_____

_____

LESSON 4·7   **Math Boxes**

---

1. A peach costs 15¢. An apple costs 12¢. Show the coins needed to buy both. Use Ⓟ, Ⓝ, Ⓓ, and Ⓠ.

2. Show 67°F.

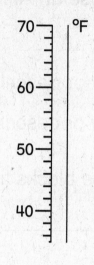

3. Write *even* or *odd* on the line.

23 _____

52 _____

258 _____

197 _____

4. Write the time.

____ : ____

5. Fill in the frames.

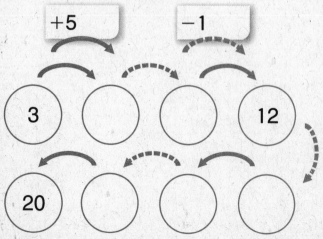

6. Use your Pattern-Block Template. Draw a hexagon.

There are _____ sides.

LESSON 4·8 Addition Practice

Write a number model to show the ballpark estimate.
Solve the problem. Show your work in the workspaces.

Unit
_____

1. Ballpark estimate:	2. Ballpark estimate:	3. Ballpark estimate:
_____	_____	_____
39 + 26	18 + 45	52 + 28

4. Ballpark estimate:	5. Ballpark estimate:	**Try This**
_____	_____	6. Ballpark estimate:
		_____
54 + 79	115 + 32	327 + 146

Add. In each problem, use the first sum to help you find the other two sums.

7. $17 + 8 =$ _____

$17 + 8 + 25 =$ _____

$17 + 8 + 25 + 12 =$ _____

8.
$$\begin{array}{r} 15 \\ + 9 \\ \hline \end{array} \quad \begin{array}{r} 15 \\ 9 \\ + 6 \\ \hline \end{array} \quad \begin{array}{r} 15 \\ 9 \\ 6 \\ + 22 \\ \hline \end{array}$$

9. $19 + 6 =$ _____

$19 + 6 + 5 =$ _____

$19 + 6 + 5 + 70 =$ _____

10.
$$\begin{array}{r} 24 \\ + 4 \\ \hline \end{array} \quad \begin{array}{r} 24 \\ 4 \\ + 7 \\ \hline \end{array} \quad \begin{array}{r} 24 \\ 4 \\ 7 \\ + 35 \\ \hline \end{array}$$

LESSON 4·8 **Math Boxes**

---

1. How much?

$[$10]$ $[$10]$ $[$5]$

Ⓠ Ⓓ Ⓓ Ⓓ Ⓓ

Ⓝ Ⓝ Ⓟ Ⓟ Ⓟ

$ _____

2. The temperature was 73°F. It got 13°F colder. What is the temperature now? _____°F
Fill in the diagram and write a number model.

Change

| Start | | End |

_____

116–118

3. 25 books. Bought 15 more. How many now? _____ books

Fill in the diagram and write a number model.

Change

| Start | | End |

_____

116–118

4. What time is it?

_____ : _____

What time will it be in a half hour?

_____ : _____

5. Draw a line segment 6 cm long. Underneath it, draw a line segment that is 2 cm longer.

6. Draw a rectangle. Two sides are 3 cm long and two sides are 5 cm long.

54

LESSON 4·9 Partial-Sums Addition Using Base-10 Blocks

Draw base-10 blocks and write the number sentence to solve each problem.

1. Example:

Unit

books

	Tens	Ones
23 + 46	‖ ‖‖	••• ••• •••
	60 +	9

Answer: _69 books_

2.

Unit

	Tens	Ones
41 + 35		
	___ +	___

Answer: _____

3.

Unit

	Tens	Ones
67 + 38		
	___ +	___

Answer: _____

4.

Unit

	Hundreds	Tens	Ones
123 + 128			
	___ +	___ +	___

Answer: _____

LESSON 4·9   # Addition Practice

Write a number model to show your ballpark estimate. Solve the problem. Show your work. Use the ballpark estimate to check whether your exact answer makes sense.

Unit
_____

1. Ballpark estimate:	**2.** Ballpark estimate:	**3.** Ballpark estimate:
_____	_____	_____
59 + 8	67 + 7	47 + 32
4. Ballpark estimate:	**5.** Ballpark estimate:	**6.** Ballpark estimate:
_____	_____	_____
58 + 26	122 + 53	136 + 157

LESSON 4·9   The Time of Day

For Problems 1–4, draw the hour hand and the minute hand
to show the time.

1. Luz got up at 7:00.

She had breakfast an hour later.

Show the time when she had breakfast.

2. The second graders went on a field trip.

They left school at 12:30.

They got back 2 hours later.

Show the time when they got back.

3. Tony left home at 8:15.

It took him half an hour to get to

school. Show the time when

he arrived at school.

4. Ming finished reading a story at 10:30.

It took her 15 minutes. Show the time

when she started reading.

5. The clock shows when Bob went to bed.

He went to sleep 15 minutes later.

At what time did he go to sleep?

_____ : _____

LESSON 4·9   **Math Boxes**

---

1. I bought a radio for $67.00.
I paid with $100.00.
How much change did I get?

_____

2. What is the temperature?
Circle the best answer.

A. 34°F

B. 36°F

C. 32°F

D. 4°F

---

3. Write three even and three odd 2-digit numbers. Circle the even numbers.

_____   _____

_____   _____

_____   _____

MRB 97

4. What time is it?

_____ : _____

What time will it be in 15 minutes?

_____ : _____

---

5. Find the rules.

11   15   9   13

11   7

MRB 98 99

6. Use your Pattern-Block Template. Draw the triangles.

What is the difference between the triangles?

MRB 54

LESSON 4·10 Math Boxes

1. Draw a rectangle that has two sides that are 4 cm long and two sides that are 7 cm long.

2. Measure the line segment with the cm side of your ruler.

_____

How long is it? _____ cm

3. Use your Pattern-Block Template to draw a polygon.

What is the name of the polygon you drew?

_____

4. A 6-sided die is shaped like a cube. What is the shape of one of its sides? Circle the best answer.

   A. square

   B. triangle

   C. hexagon

   D. circle

5. Draw a line segment 2 cm long.

6. Use your template. Draw the other half of the shape and write the name of it.

LESSON
5·1
Math Boxes

---

1. 843

There are _____ hundreds.

There are _____ tens.

There are _____ ones.

MRB
10

---

2. Read the tally chart. How many children in the class are 8 years old? _____

Class Ages	
Age	Number of Children
7	~~HHT~~ ~~HHT~~ I
8	~~HHT~~ I
9	III

MRB
40

---

3. How many in all? Circle the best answer.

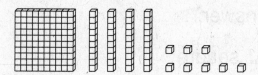

A 841   **B** 481

C 408   **D** 148

MRB
11

---

4. Fill in the missing numbers.

144

MRB
8

---

5. Draw a triangle. Make each side 4 cm long.

---

6. Graph this data. Jamar earned 3 stickers on Monday, 2 on Tuesday, and 4 on Wednesday.

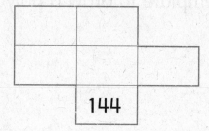

Jamar's Sticker Graph

Number of Stickers

M   Tu   W   Th   F
Day of the Week

LESSON 5·2

Using Secret Codes to Draw Shapes

1. The codes show how to connect the points with line segments. Can you figure out how each code works? Discuss it with your partner.

Code:
A→E→D

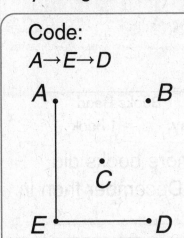

Code:
E→A→C→B→D

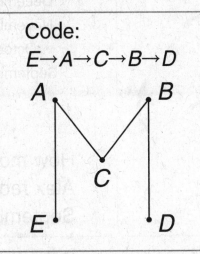

Code:
E→A→C→D→B

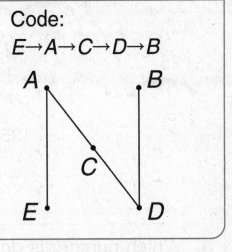

2. Use each code. Draw line segments using your straightedge.

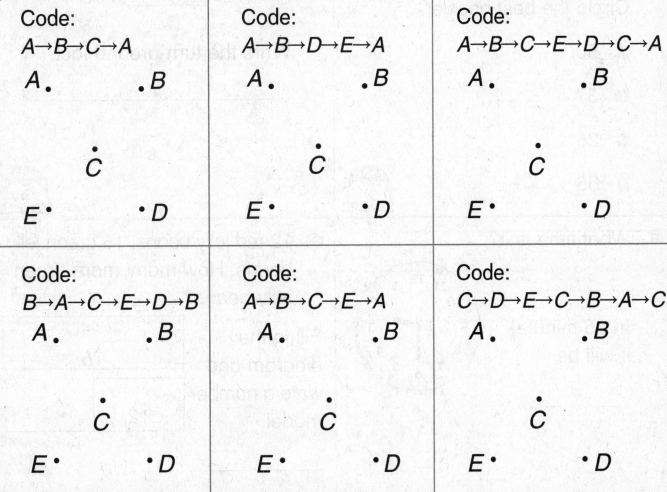

Code:
A→B→C→A

A.   .B

C

E•   •D

Code:
A→B→D→E→A

A.   .B

C

E•   •D

Code:
A→B→C→E→D→C→A

A.   .B

C

E•   •D

Code:
B→A→C→E→D→B

A.   .B

C

E•   •D

Code:
A→B→C→E→A

A.   .B

C

E•   •D

Code:
C→D→E→C→B→A→C

A.   .B

C

E•   •D

LESSON
5·2

Math Boxes

---

1. Fill in the missing numbers.

Rule
+25

50

125

2.

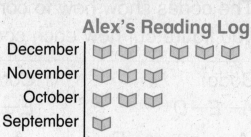

Alex's Reading Log

December

November

October

September

Books Read

Key: ☐ = 1 book

How many more books did Alex read in December than in September? _____

---

3. Which number is an odd number less than 50? Circle the best answer.

A 50

B 37

C 26

D 55

MRB
97

---

4. Solve.

Unit

8 + 6 = _____

Write the turn-around fact.

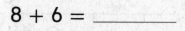

MRB
22

---

5. What time is it?

____:____

In 15 minutes it will be

____:____.

---

6. 12 red jelly beans. 16 green jelly beans. How many more green jelly beans? _____

Fill in the diagram and write a number model.

Quantity
16

Quantity
12

_____
Difference

_____

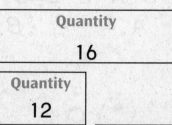

MRB
110 111

LESSON 5·3 Parallel or Not Parallel?

These line segments are **parallel**.

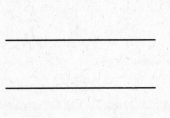

These line segments are **not parallel**.

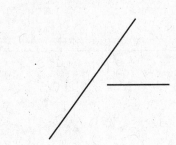

Quadrangles (Quadrilaterals)

These polygons are quadrangles (quadrilaterals).

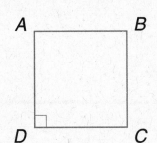

square

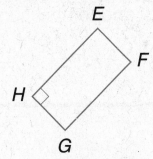

rectangle

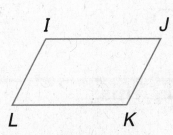

parallelogram

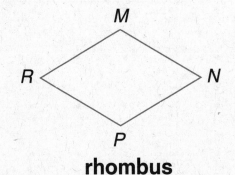

rhombus

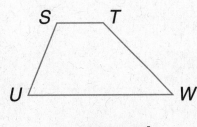

trapezoid

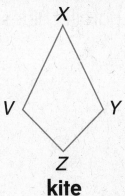

kite

LESSON 5·3   Parallel Line Segments

Use a straightedge.

1. Draw line segments *AB* and *CD*.

   Are line segments *AB* and *CD* parallel?

   _____

   A • • • • B
   • • • • •
   C • • • D
     E   F
   • • • • •

2. Draw line segment *EF*.

   Are line segments *AB* and *EF* parallel?

   _____

3. Draw line segment *LM*.

4. Draw a line segment that is parallel
   to line segment *LM*. Label its endpoints
   R and *S*.

5. Draw a line segment that is NOT parallel
   to line segment *LM*. Label its endpoints
   T and *U*.

   L • • • • •
   • • • • •
   • • • • •
   • • • • •
             M

Try This

6. Draw a quadrangle that has
   NO parallel sides.

   • • • • •
   • • • • •
   • • • • •
   • • • • •
   • • • • •

**LESSON
5·3**   **Parallel Line Segments** *continued*

7. Draw a quadrangle in which opposite
   sides are parallel.

8. Draw a quadrangle in which all
   four sides are the same length.

   What is another name for this shape?

   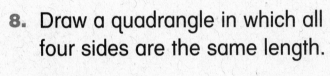

9. Draw a quadrangle in which 2 opposite
   sides are parallel and the other
   2 opposite sides are NOT parallel.

LESSON 5·3

Open Number Lines

You can use an open number line to add.

Example: Solve 34 + 23.

◆ Draw a line. Make and label point 34.

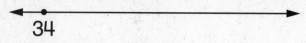

34

◆ Think 23 = 2 tens and 3 ones.

◆ Start at 34 and count up 1 ten. Make a point at 44. Count up one more ten. Make a point at 54.

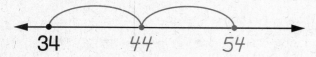

34   44   54

◆ Count up 3 ones. Make a point at 57.

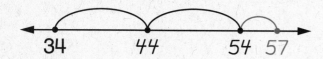

34   44   54   57

Answer: 34 + 23 = **57**

Use an open number line to add.

1. 43 + 17

2. 56 + 25

43

56

Answer: 43 + 17 = _____

Answer: 56 + 25 = _____

Math Boxes

---

1. Use the digits 1, 3, and 5 to make:

the smallest number possible.

_____

the largest number possible.

_____

---

2. On which day did Ms. Forbes' class have the most recess minutes? _____

Ms. Forbes' Class Recess Minutes	
Day	Number of Minutes
Wednesday	//// //// //// ////
Thursday	//// //// //// //// ////
Friday	//// //// ////

---

3. How many in all?

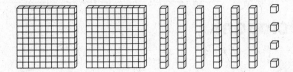

_____

MRB 11

---

4. Complete the number grid.

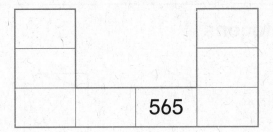

565

MRB 8

---

5. Draw a square. Make each side 3 cm long.

---

6. Graph this data.

On Tuesday it was 20°C, on Wednesday it was 35°C, and on Thursday it was 30°C.

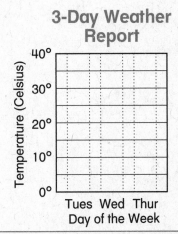

3-Day Weather Report

LESSON 5·4   Polygons

Triangles

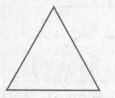 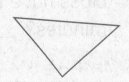

Quadrangles or Quadrilaterals

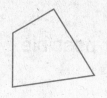

Pentagons

  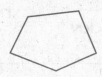

Hexagons

Heptagons

Octagons

These are NOT polygons.

 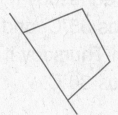 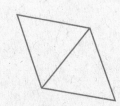

LESSON 5·4   Math Boxes

1. Fill in the missing numbers.

Rule

−10¢

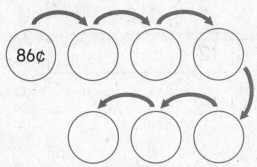

86¢

2. How many patients did Dr. Rios see on Tuesday? ____

Patient Log

Key:
☺ = 1 patient

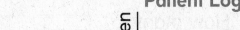

Patients Seen

M	Tu	W

Day of the Week

3. Write 5 even numbers greater than 50.

MRB
97

4. Solve.

$9 + 3 =$ _____

Unit

cats

Write the turn-around fact.

_____

MRB
22

5. Draw the hands to show 6:45.

$\frac{1}{2}$ hour earlier is ____:____.

6. 35 butterflies. 10 flew away. How many butterflies are left?

_____ butterflies

Fill in the diagram and write a number model.

Quantity
35

Quantity
10

Difference

_____

MRB
110 111

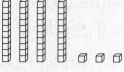

Math Boxes

1.

How many? _____

Cross out 13 cubes.

How many are left? _____

Write a number model.

_____ − _____ = _____

11

2. Fill in the missing numbers.

	112

121	

8

3. Use your ruler. Draw $\overline{MS}$.

$\overline{MS}$ is _____ cm long.

50 66

4. A triangle has _____ sides.

A rhombus has _____ sides.

A trapezoid has _____ sides.

A hexagon has _____ sides.

54 55

5. How many dots are in this 4-by-6 array? Count by 4s.

• • • • • •
• • • • • •
• • • • • •
• • • • • •

_____ dots

6. Write 3 numbers that add up to 20.

___ + ___ + ___ = ___

LESSON 5·6 Connecting Points

Draw a line segment between each pair of points.
Record how many line segments you drew.

Example:

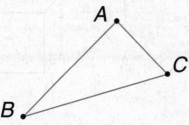

3 points

_3_ line segments

1.

P.

A.

.L

U .

4 points

_____ line segments

2.

R.

O.

•E

S.

•I

5 points

_____ line segments

Try This

3. Connect the points in order from 1 to 3. Use a straightedge.

Find and name 3 triangles.

_____

Try to name a fourth triangle.

_____

Color a 4-sided figure.

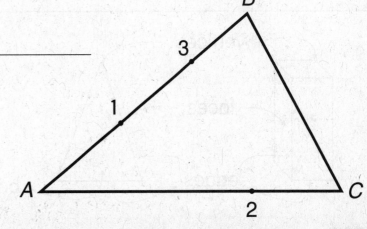

LESSON 5·6   3-D Shapes Poster

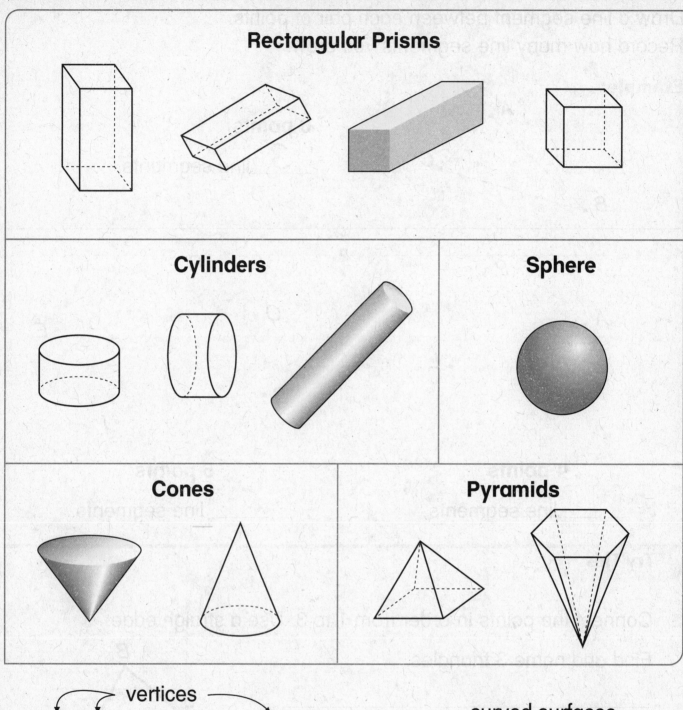

Rectangular Prisms

Cylinders

Sphere

Cones

Pyramids

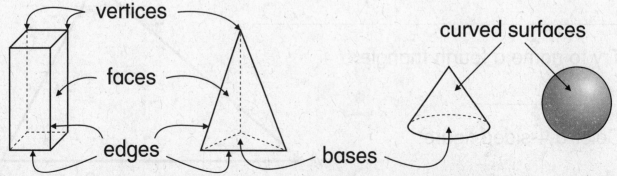

vertices

faces

edges

bases

curved surfaces

LESSON 5·6   # What's the Shape?

Write the name of each shape.

1.

_____

2.

_____

3.

_____

4.

_____

5.
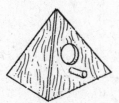
_____

6.

_____

7.
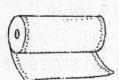
_____

8.

_____

9.
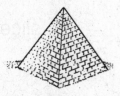
_____

10.

_____

LESSON 5·6 Math Boxes

1. Count by 1,000.

1,300; _____; _____;

_____; _____; _____;

_____; _____

2. Match quarters to ¢.

5 quarters   150¢

6 quarters   250¢

7 quarters   125¢

10 quarters   175¢

3. **Ways to Get to School**

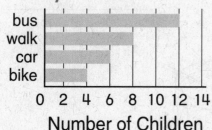

Number of Children

How many children walk to school? _____

4. Solve.

$3 + 5 =$ _____

$30 + 50 =$ _____

$300 + 500 =$ _____

_____ $= 6 + 8$

_____ $= 60 + 80$

_____ $= 600 + 800$

Unit

5. Buy a model dinosaur for 37¢. Pay with 2 quarters.

How much change? _____

6. 4 children share 12 slices of pizza equally. How many slices does each child get? Draw a picture.

Each child gets _____ slices.

LESSON 5·7 Math Boxes

1.

How many? _____

Cross out 14 cubes.

How many are left? _____

Write a number model.

_____ – _____ = _____

2. Complete the number grid.

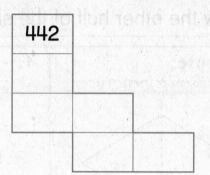

MRB
8

3. Use your ruler. Draw line segment *AB*.

$\overline{AB}$ is _____ cm long.

MRB
50 66

4. This is a trapezoid. Put an X on the line segments that are parallel.

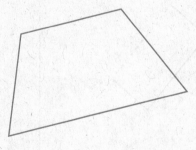

MRB
51

5. How many dots are in this 5-by-4 array? Count by 5s. Circle the best answer.

 _____ dots in all.

A 19 **B** 20

C 25 **D** 15

6. Write 3 names for 15.

_____ + _____ + _____ = 15

15 = _____ + _____ + _____

_____ + 9 + _____ = 15

LESSON 5·8 Symmetrical Shapes

Each picture below shows half of a shape on your Pattern-Block Template. Guess what the full shape is. Then use your template to draw the other half of the shape. Write the name of the shape.

Example:

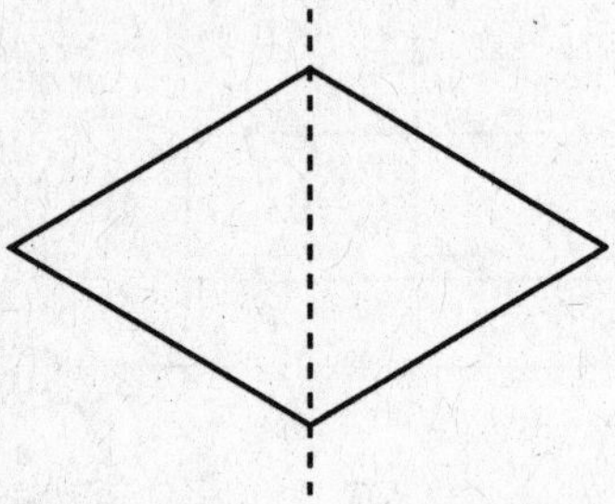

rhombus

1.

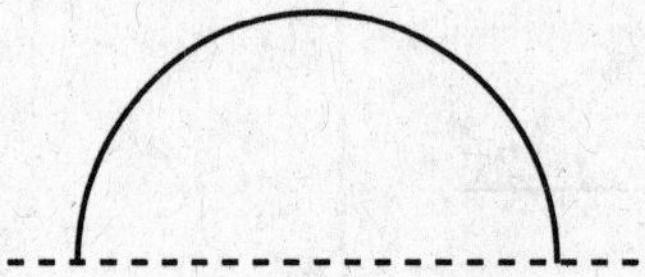

_____

2.

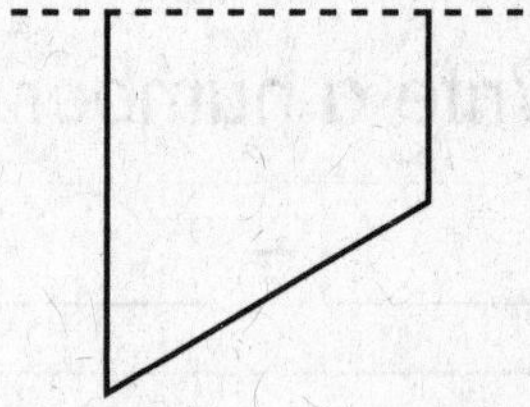

_____

3.

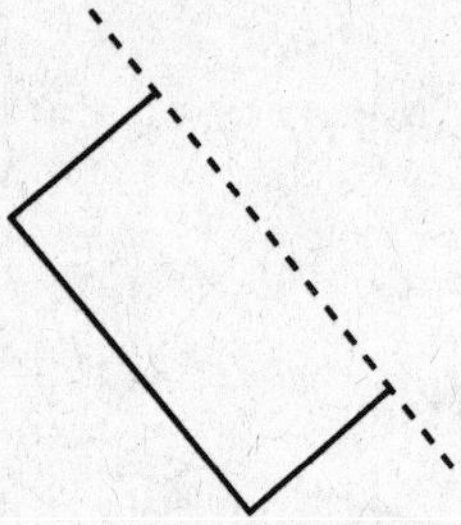

_____

4.

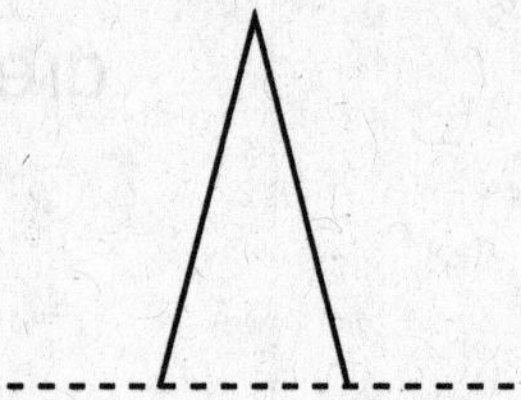

_____

5.

_____

6.

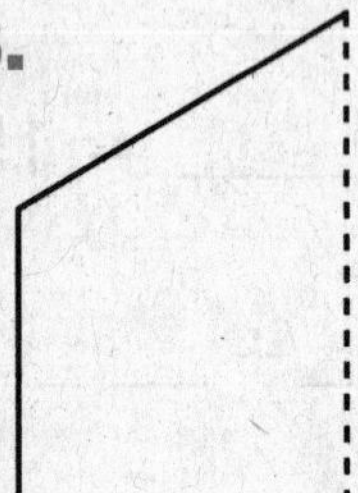

_____

7.

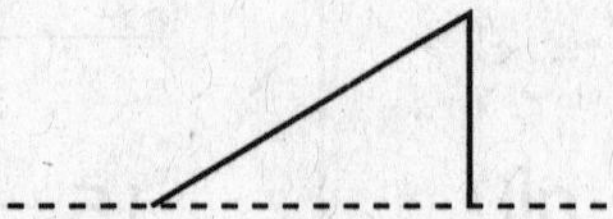

_____

8.

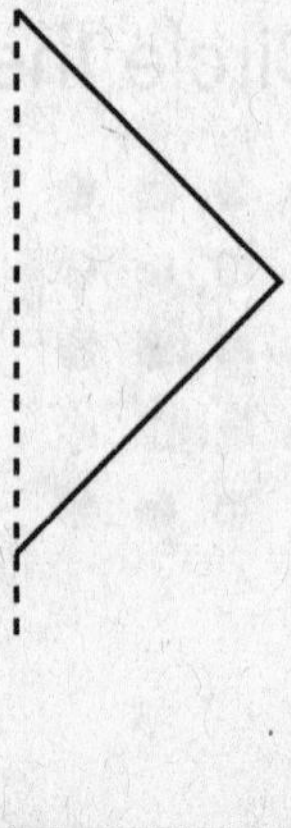

_____

LESSON 5·8 Math Boxes

---

1. Count by 1,000.

2,600; _____; _____;

_____; _____; _____;

_____; _____

---

2. Write <, >, or =.

4 dimes _____ 50¢

3 quarters _____ 75¢

$1.00 _____ 11 dimes

3.

Favorite Sport

Number of Children

How many more children like basketball than soccer?

_____

---

4. Solve.

18 − 9 = _____

180 − 90 = _____

1,800 − 900 = _____

Unit

---

5. Spend 68¢. Pay with $1.00. How much change?

_____

---

6. Share 19¢ equally among 5 children. How many cents does each child get?_____

How many cents are left over? Circle the best answer.

A 0¢   **B** 5¢

C 4¢   **D** 2¢

LESSON
5·9

Math Boxes

---

1. Draw a 7-by-9 array.

---

2. How many hot dogs were sold in the 2nd hour of the game? _____

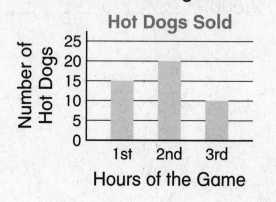

Hot Dogs Sold

---

3. Graph this data.

On a nature walk Amy saw 5 snails, 2 turtles, and 6 caterpillars.

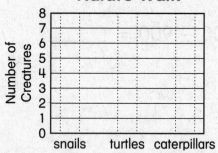

Nature Walk

---

4. Three children share 18 pieces of paper equally.

How many pieces of paper will each child get? _____

How many pieces of paper will be left over? _____

---

5. How many dots are in this 4-by-4 array?

• • • •
• • • •
• • • •
• • • •

_____

---

6. 28 soccer balls. 36 basketballs. How many more basketballs?

_____ basketballs

Fill in the diagram and write a number model.

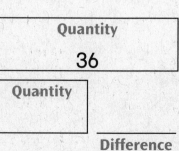

Quantity
36

Quantity

_____
Difference

**LESSON
6·1**   *Three Addends* **Record Sheet**

For each turn:

◆ Write the 3 numbers.

◆ Add the numbers.

◆ Write a number model to show the order in which
   you added.

1. Numbers: ____, ____, ____

   Number model:

   ____ + ____ + ____ = ____

2. Numbers: ____, ____, ____

   Number model:

   ____ + ____ + ____ = ____

3. Numbers: ____, ____, ____

   Number model:

   ____ + ____ + ____ = ____

4. Numbers: ____, ____, ____

   Number model:

   ____ + ____ + ____ = ____

5. Numbers: ____, ____, ____

   Number model:

   ____ + ____ + ____ = ____

6. Numbers: ____, ____, ____

   Number model:

   ____ + ____ + ____ = ____

For Problems 7 and 8, use 4 numbers.

7. Numbers: ____, ____, ____, ____

   Number model: ____ + ____ + ____ + ____ = ____

8. Numbers: ____, ____, ____, ____

   Number model: ____ + ____ + ____ + ____ = ____

LESSON 6·1 Ballpark Estimates

Fill in the unit box. Then, for each problem:

Unit

◆ Make a ballpark estimate.

◆ Write a number model for your estimate. Then solve the problem. For Problems 1, 2, and 3 use the Partial-Sums Algorithm. For Problems 4–6, use any strategy you choose.

◆ Compare your estimate to your answer.

1. Ballpark estimate: _____ $29 + 7 =$ _____	**2.** Ballpark estimate: _____ $87 + 9 =$ _____	**3.** Ballpark estimate: _____ $37 + 42 =$ _____
4. Ballpark estimate: _____ $27 + 13 =$ _____	**5.** Ballpark estimate: _____ $138 + 46 =$ _____	**6.** Ballpark estimate: _____ $142 + 128 =$ _____

LESSON 6·1

Math Boxes

---

1. Which is least likely to happen? Choose the best answer.

⬭ A dog will have puppies today.

⬭ You will go to sleep tonight.

⬭ Chocolate milk will come out of the water fountain.

⬭ Someone will read you a book.

---

2. How long is this line segment?

_____

about _____ cm

---

3. Arlie harvested 12 bushels of corn and 19 bushels of tomatoes. How many bushels in all? _____ bushels

Fill in the diagram and write a number model.

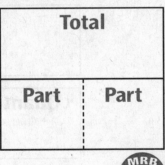

Total	
Part	Part

MRB 109

_____

---

4. How many dots in this 2-by-8 array?

• • • • • • • •
• • • • • • • •

_____

---

5. Put these numbers in order from least to greatest and circle the middle number (the median).

109, 99, 129

---

6. Use your calculator. Start at 92. Count by 5s.

92, 97, _____, _____, _____,

_____, _____, _____, _____

MRB 6

Comparison Number Stories

For each number story:

◆ Write the numbers you know in the comparison diagram.

◆ Write ? for the number you want to find.

◆ Solve the problem.

◆ Write a number model.

---

1. Barb scored 27 points.
Cindy scored 10 points.

Barb scored _____ more points
than Cindy.

Number model: _____

Quantity

Quantity

Difference

---

2. Frisky lives on the 16th floor.
Fido lives on the 7th floor.

Frisky lives _____ floors higher
than Fido.

Number model: _____

Quantity

Quantity

Difference

---

3. Ida is 36 years old. Bob is 20 years old.

Ida is _____ years older than Bob.

Number model: _____

Quantity

Quantity

Difference

LESSON 6·2

Comparison Number Stories *continued*

4. A jacket costs $75.
   Pants cost $20.

   The pants cost $ _____ less
   than the jacket.

   Number model: _____

Quantity

Quantity

   _____
   Difference

Try This

5. Jack scored 13 points. Jack scored 6
   more points than Eli.

   Eli scored _____ points.

   Number model: _____

   or _____

Quantity

Quantity

   _____
   Difference

6. Billy is 16 years old. Paul is
   6 years younger than Billy.

   Paul is _____ years old.

   Number model: _____

   or _____

Quantity

Quantity

   _____
   Difference

7. Marcie is 56 inches tall.
   Nick is 70 inches tall.

   Marcie is _____ inches shorter than Nick.

   Number model:

   or

Quantity

Quantity

   _____
   Difference

LESSON
6·2   **Math Boxes**

---

1. Fill in the frames.

MRB
98 99

---

2. Draw all the possible ways to show 30¢ using Ⓠ, Ⓓ, Ⓝ.

---

3. Write <, >, or =.

Unit

7 + 5 + 30 _____ 40

11 + 6 + 4 _____ 26

32 _____ 18 + 7 + 2

19 _____ 13 + 9 + 1

MRB
9

---

4. Draw a line segment that is about 5 cm long.

---

5. Write the time.

_____ : _____

---

6. Halve.

2 _____   10 _____

4 _____   14 _____

8 _____   50 _____

LESSON 6·3

What Is Your Favorite Food?

1. Make tally marks to show the number of children who chose a favorite food in each group.

fruit/ vegetables	bread/cereal/ rice/pasta	dairy products	meat/poultry/fish/ beans/eggs/nuts

2. Make a graph that shows how many children chose a favorite food in each group.

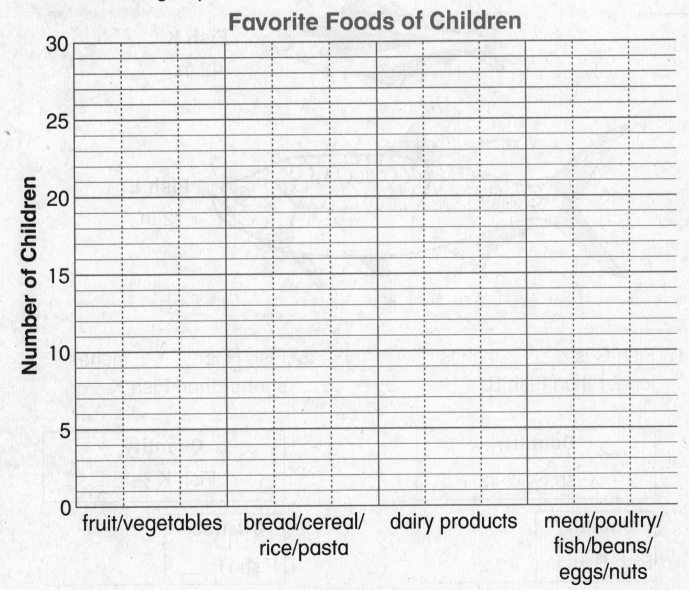

Favorite Foods of Children

Food Groups

LESSON 6·3 Comparing Fish

Fish Lengths

 Fish B
14 in.

 **Fish C**
18 in.

 **Fish D**
24 in.

 **Fish H**
30 in.

 **Fish K**
42 in.

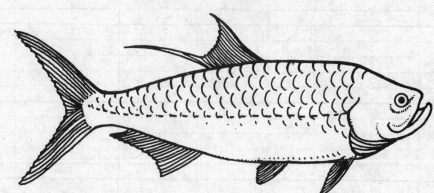 **Fish L**
72 in.

1. Fish C is _____ inches longer than Fish B.

2. Fish H is _____ inches shorter than Fish K.

Quantity
Fish C 18 in.

Quantity
Fish B 14 in.

_____
Difference

Quantity
Fish K _____

Quantity
Fish H _____

_____
Difference

LESSON 6·3

Comparing Fish *continued*

3. Fish L is _____ inches longer than Fish H.

Quantity

Quantity

_____
Difference

4. Fish B is _____ inches shorter than Fish D.

Quantity

Quantity

_____
Difference

5. Fish H is 6 inches longer than _____.

Quantity

Quantity

_____
Difference

6. Fish L is 30 inches longer than _____.

Quantity

Quantity

_____
Difference

7. Fish C is 6 inches shorter than _____.

Quantity

Quantity

_____
Difference

8. Fish B is 16 inches shorter than _____.

Quantity

Quantity

_____
Difference

Math Boxes

1. Which of these is unlikely to happen? Circle the sentence.

It is unlikely that...

you will have lunch today.

you will grow 5 inches while you sleep tonight.

you will see someone you recognize.

2. Measure the line segment.

_____

about _____ in.

about _____ cm

3. Denise picked 16 tulips and 4 daffodils. How many flowers did she pick in all?

_____ flowers

Fill in the diagram and write a number model.

Total	
Part	**Part**

_____

109

4. How many dots are in this 3-by-5 array?

• • • • •
• • • • •
• • • • •

_____

5. List the numbers in order.
5, 10, 4, 6, 4, 4, 6

_____

Which number occurs most

often? _____

6. Continue.

312, 314, 316, _____, _____,

_____, _____, _____,

_____

LESSON 6·4

Addition and Subtraction Number Stories

Do the following for each problem:

◆ Read the number story.

◆ Solve the problem. You may use the diagrams on *Math Masters,* page 437.

◆ Write the answer and a number model.

1. Rushing Waters has 26 water slides. Last year, there were only 16 water slides. How many new slides are there this year?

   There are _____ new water slides.

   Number model:

   _____

2. On Monday, Mary had 30 shells. On Tuesday, her aunt gave her 12 more shells. On Wednesday, Mary gave 7 shells to her friend. How many shells does Mary have now?

   Mary has _____ shells.

   Number model:

   _____

3. Yvette bought 25 red balloons and 25 white balloons for a party. During the party, 8 of the balloons popped. How many balloons did she have when the party ended?

   Yvette had _____ balloons.

   Number model:

   _____

LESSON 6·4 Ballpark Estimates

Fill in the unit box. Then, for each problem:

Unit

◆ Make a ballpark estimate before you add.

◆ Write a number model for your estimate.

◆ Use your calculator to solve the problem. Write your answer in the answer box.

◆ Compare your estimate to your answer.

1. Ballpark estimate: _____ 48 + 7 **Answer**	**2.** Ballpark estimate: _____ 63 + 9 **Answer**	**3.** Ballpark estimate: _____ 33 + 45 **Answer**
4. Ballpark estimate: _____ 23 + 14 **Answer**	**5.** Ballpark estimate: _____ 248 + 132 **Answer**	**6.** Ballpark estimate: _____ 445 + 238 **Answer**

LESSON 6·4 Math Boxes

1. Fill in the frames.

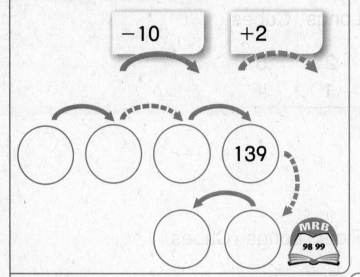

−10   +2

139

98 99

2. Draw all the possible ways to show 35¢ using Ⓠ, Ⓓ, Ⓝ.

MRB
88 89

3. Write >, <, or =.

Unit

5 + 3 + 18 _____ 20

8 + 7 + 5 _____ 36

76 _____ 28 + 32 + 21

100 _____ 25 + 25 + 50

MRB 9

4. Which line segment is about 1 inch long? Choose the best answer.

⬭ ————————————————

⬭ ——————————

⬭ ——————

⬭ ———

5.

1 hour earlier is ____:____.

1 hour later is ____:____.

6.

in	out
4	
8	
20	
	58

Rule
halve

101 102

LESSON 6·5 Subtraction

Use base-10 blocks to help you subtract.

1.
Longs	Cubes
1	9
−	7

2.
Longs	Cubes
2	5
− 1	4

3.
Longs	Cubes
4	3
− 1	8

4.
Flats	Longs	Cubes
1	3	6
−	4	7

Use any strategy to solve.

5.   14
    −   6

6.   38
    − 23

7.   124
    − 26

8.   164
    − 126

Math Boxes

---

1. Circle the one that is likely to happen.

It is likely that...

you will do a Math Box today.

you will fly like a bird.

an elephant will visit the classroom.

---

2. Measure the line segment.

_____

about _____ in.

about _____ cm

---

3. Kurtis scored 13 points in the first half of the game and a total of 24 points by the end. How many points did Kurtis score in the second half? _____ points

Total	
Part	**Part**

Number model:

_____

---

4. How many dots are in this 7-by-9 array?

_____

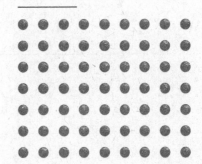

---

5. Which number occurs most often? Choose the best answer.

8, 17, 9, 8, 10

⬭ 9

⬭ 17

⬭ 10

⬭ 8

---

6. Use your calculator. Count by 9s. Start at 76.

76, _____, _____, _____,

_____, _____

What pattern do you see?

_____

_____

LESSON 6·6 How Many Children Get *n* Things?

Follow the directions on *Math Masters,* page 176 to fill in the table.

What is the total number of counters?	How many counters are in each group? (Roll a die to find out.)	How many groups are there?	How many counters are left over?

LESSON 6·6   **Making a Picture Graph**

A zoo has a collection of birds. One month, they recorded the number of eggs the birds laid. The number of eggs a bird lays at one time is called a *clutch.* Use the information in the table below to create a picture graph.

Type of Bird	American Robin	Canada Goose	Flamingo	Mallard Duck	Toucan
Number of Eggs in a Clutch	6	10	1	12	4

Bird Eggs Laid in 1 Month

Number of Eggs in a Clutch

	American Robin	Canada Goose	Flamingo	Mallard Duck	Toucan
12					
11					
10					
9					
8					
7					
6					
5					
4					
3					
2					
1					

KEY: ◯ = 1 egg

LESSON 6·6 Making a Picture Graph *continued*

Use the information from the picture graph on journal page 146A to solve the number stories.

1. The American Robin and Toucan are tree birds. What is the total number of eggs laid by the tree birds?

   Answer: _____ eggs

   Number model: _____

2. The Canada Goose, Flamingo, and Mallard Duck are water birds. How many eggs did the water birds lay in total?

   Answer: _____ eggs

   Number model: _____

3. How many more eggs did the Mallard Duck lay than the American Robin?

   Answer: _____ eggs

   Number model: _____

4. Use the information from the graph to write your own number story. Have a partner solve the problem.

   Answer: _____

   Number model: _____

LESSON 6·6 Math Boxes

1. 639 has

_____ hundreds

_____ tens

_____ ones

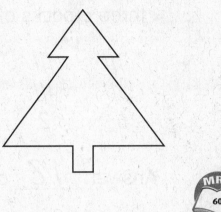 MRB 10 11

2. Draw the line of symmetry.

MRB 60

3. Share 1 dozen cookies equally among 5 children. Draw a picture.

Each child gets _____ cookies.

There are _____ cookies left over.

4. Use counters to make a 5-by-2 array. Draw the array.

How many counters in all?

_____ counters

5. The temperature is _____ °F.

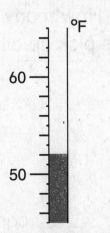

6. Fill in the pattern.

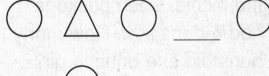

LESSON 6·7 Multiplication Stories

Solve each problem. Draw pictures or use counters to help.

Example: How many cans are in three 6-packs of juice?

/// /// ///
/// /// ///
6   12   18

Answer: __18__ cans

---

1. Mr. Yung has 4 boxes of markers. There are 6 markers in each box. How many markers does he have in all?

Answer: _____ markers

2. Sandi has 3 bags of marbles. Each bag has 7 marbles in it. How many marbles does she have in all?

Answer: _____ marbles

---

3. Mrs. Jayne brought 5 packages of buns to the picnic. Each package had 6 buns in it. How many buns did she bring in all?

Answer: _____ buns

4. After the picnic, 5 boys each picked up 4 soft-drink cans to recycle. How many cans did the boys pick up all together?

Answer: _____ cans

Math Boxes

---

1. Write the number.

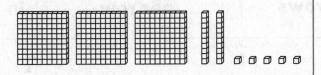

_____

2. Make ballpark estimates. Write a number model for each estimate.

Unit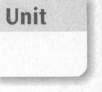

32 + 59

_____ + _____ = _____

51 + 27

_____ + _____ = _____

MRB
92–94

MRB
10 11

---

3. Measure the line segment.

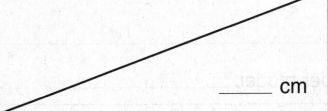

_____ cm

Draw a line segment 3 cm shorter.

4.

How many? _____

Cross out 23 cubes.

How many are left? _____

Write a number model.

_____ – _____ = _____

MRB
31

---

5. Lauren's birthday is on the tenth day in the shortest month of the year. In what month is her birthday?

MRB
85

6. Double.

Unit

2 _____

4 _____

10 _____

50 _____

LESSON 6·8 Array Number Stories

Array

o o o o o o o o o o
o o o o o o o o o o
o o o o o o o o o o
o o o o o o o o o o
o o o o o o o o o o
o o o o o o o o o o

Multiplication Diagram

rows	_____ per row	_____ in all

Number model: _____

Array

o o o o o o o o o o
o o o o o o o o o o
o o o o o o o o o o
o o o o o o o o o o
o o o o o o o o o o
o o o o o o o o o o

Multiplication Diagram

rows	_____ per row	_____ in all

Number model: _____

Array

o o o o o o o o o o
o o o o o o o o o o
o o o o o o o o o o
o o o o o o o o o o
o o o o o o o o o o
o o o o o o o o o o

Multiplication Diagram

rows	_____ per row	_____ in all

Number model: _____

Array

o o o o o o o o o o
o o o o o o o o o o
o o o o o o o o o o
o o o o o o o o o o
o o o o o o o o o o
o o o o o o o o o o

Multiplication Diagram

rows	_____ per row	_____ in all

Number model: _____

Multiplication Number Stories

LESSON 6·8

For each problem:
- ◆ Use Xs to show the array.
- ◆ Answer the question.
- ◆ Write a number model.

1. The marching band has 3 rows with 5 players in each row. How many players are in the band?

○ ○ ○ ○ ○ ○ ○ ○ ○ ○
○ ○ ○ ○ ○ ○ ○ ○ ○ ○
○ ○ ○ ○ ○ ○ ○ ○ ○ ○
○ ○ ○ ○ ○ ○ ○ ○ ○ ○
○ ○ ○ ○ ○ ○ ○ ○ ○ ○
○ ○ ○ ○ ○ ○ ○ ○ ○ ○

There are _____ players.
Number model:

_____

2. Mel folded his paper into 2 rows of 4 boxes each. How many boxes did he make?

○ ○ ○ ○ ○ ○ ○ ○ ○ ○
○ ○ ○ ○ ○ ○ ○ ○ ○ ○
○ ○ ○ ○ ○ ○ ○ ○ ○ ○
○ ○ ○ ○ ○ ○ ○ ○ ○ ○
○ ○ ○ ○ ○ ○ ○ ○ ○ ○
○ ○ ○ ○ ○ ○ ○ ○ ○ ○

He made ____ boxes.
Number model:

_____

3. The sheet has 5 rows of stamps. There are 5 stamps in each row. How many stamps are there?

○ ○ ○ ○ ○ ○ ○ ○ ○ ○
○ ○ ○ ○ ○ ○ ○ ○ ○ ○
○ ○ ○ ○ ○ ○ ○ ○ ○ ○
○ ○ ○ ○ ○ ○ ○ ○ ○ ○
○ ○ ○ ○ ○ ○ ○ ○ ○ ○
○ ○ ○ ○ ○ ○ ○ ○ ○ ○

There are ____ stamps in all.
Number model:

_____

4. The orchard has 4 rows of trees. Each row has 8 trees. How many trees are there?

○ ○ ○ ○ ○ ○ ○ ○ ○ ○
○ ○ ○ ○ ○ ○ ○ ○ ○ ○
○ ○ ○ ○ ○ ○ ○ ○ ○ ○
○ ○ ○ ○ ○ ○ ○ ○ ○ ○
○ ○ ○ ○ ○ ○ ○ ○ ○ ○
○ ○ ○ ○ ○ ○ ○ ○ ○ ○

There are _____ trees.
Number model:

_____

LESSON 6·8   Math Boxes

1. Choose the best answer. Be careful!

6 tens

3 ones

8 hundreds

○ 638   ○ 836

○ 368   ○ 863

2. Draw the line of symmetry.

3. Use counters to solve.

$14.00 is shared equally.

Each child gets $5.00.

How many children are sharing?

_____ children

How many dollars are left over?

_____ dollars

4. This is a _____-by-_____ array.

How many dots in all?

_____ dots

5. Find the differences.

16°C and 28°C _____

70°F and 57°F _____

15°C and 43°C _____

6. What comes next?

   _____

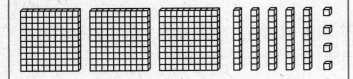

LESSON 6·9 **Math Boxes**

---

1. How much?

_____

MRB
11

2. Make a ballpark estimate.

Write a number model for your estimate.

Unit

$49 + 51$

Number Model:

_____

MRB
92–94

---

3. Draw a line segment that is about 3 inches long.

Now draw a line segment that is 1 inch shorter.

How long is this line segment? Choose the best answer.

⬭ 4 inches ⬭ 2 inches

⬭ 1 inch ⬭ 5 inches

4. There are 22 first graders and 35 second graders. How many more second graders?

_____ more

| Quantity |
| |

Fill in the diagram and write a number model.

| Quantity |
| |

_____ _____

_____

MRB
110 111

---

5. Write <, >, or =.

1 week _____ 7 days

48 hours _____ 1 day

6 months _____ 1 year

MRB
9

6.

Rule

double

in	out
2	
3	
	30
100	

LESSON 6·9   *Array Bingo* **Directions**

Materials
- ☐ 2 six-sided dice, 1 twelve-sided die, or an egg-carton number generator

- ☐ 9 cards labeled "A" cut from *Math Masters,* p. 450 for each player

Players   2–5

Skill   Recognize an array for a given number.

Object of the Game   Turn over a row, column, or diagonal of cards.

Directions

1. Each player arranges the 9 cards at random in a 3-by-3 array.

2. Players take turns. When it is your turn:

   Generate a number from 1 to 12, using the dice, die, or number generator. This number represents the total number of dots in an array.

   Look for the array card with that number of dots. Turn that card facedown.

3. The first player to have a row, column, or diagonal of facedown cards calls "Bingo!" and wins the game.

Array Bingo Directions *continued*

Another Way to Play

Materials
- ☐ 1 twenty-sided die or number cards with one card for each of the numbers 1–20

- ☐ all 16 array cards from *Math Masters,* p. 450 for each player

Each player arranges his or her cards at random in a 4-by-4 array. Players generate numbers using the die or number cards. If you use number cards to generate numbers, do this:

- ◆ Shuffle the cards.

- ◆ Place them facedown on the table.

- ◆ Turn over the top card.

- ◆ If all 20 cards are turned over before someone calls "Bingo," reshuffle the deck and use it as before.

LESSON 6·10 Division Problems

Use counters or simple drawings to find the answers. Fill in the blanks.

1. 16 cents shared equally

by 2 people   by 3 people

_____ ¢ per person   _____ ¢ per person

_____ ¢ remaining   _____ ¢ remaining

by 4 people   by 5 people

_____ ¢ per person   _____ ¢ per person

_____ ¢ remaining   _____ ¢ remaining

2. 25 cents shared equally

by 3 people   by 4 people

_____ ¢ per person   _____ ¢ per person

_____ ¢ remaining   _____ ¢ remaining

by 5 people   by 6 people

_____ ¢ per person   _____ ¢ per person

_____ ¢ remaining   _____ ¢ remaining

3. 16 crayons, 6 crayons per box

How many boxes? _____   How many crayons remaining? _____

4. 24 eggs, 6 eggs in each cake

How many cakes? _____   How many eggs remaining? _____

LESSON 6·10   **Math Boxes**

---

1. Use the digits 3, 1, and 5.

Write the smallest possible number.

_____

Write the largest possible number.

_____

---

2. Do these objects have at least one line of symmetry?

Yes or No?

_____   _____

MRB
60

---

3. 18 cans of juice are shared by 5 people. Draw a picture.

_____ cans per person

_____ cans left over

---

4. 4 rows of 4 chairs. How many chairs in all? _____ chairs

Draw an array to solve. Fill in the multiplication diagram.

rows	per row	in all

---

5. Find the differences.

32°F and 53°F _____

37°C and 19°C _____

75°F and 93°F _____

---

6. Which figure does not belong? Choose the best answer.

⬭ circle with diagonal lines

⬭ empty circle

⬭ circle with vertical lines

⬭ circle with horizontal lines

LESSON 6·11   **Math Boxes**

---

1. Find the rule. Complete the table.

Rule: _____

in	out
10	5
16	8
20	
	20

MRB 100–102

---

2. How much did Kristle weigh when she was 7 years old?

_____ pounds

Kristle's Weight Chart

Weight in Pounds: 30, 40, 50, 60, 70
Age in Years: 6, 7, 8

MRB 40 41

---

3. Which number from this list occurs most often?

_____

3, 13, 23, 9, 14, 9

MRB 45

---

4. Double.

25¢ _____

50¢ _____

15¢ _____

75¢ _____

---

5. Draw the shapes that come next.

__ __ __

---

6. Count back by 10s.

220, 210, _____, _____,

_____, _____, _____,

_____, _____, _____,

_____, _____, _____

Table of Equivalencies

Weight

kilogram	1,000 g
pound	16 oz
ton	2,000 lb

1 ounce is about 30 g

<	is less than
>	is more than
=	is equal to
=	is the same as

Length

kilometer	1,000 m
meter	100 cm or 10 dm
decimeter	10 cm
centimeter	10 mm
foot	12 in.
yard	3 ft or 36 in.
mile	5,280 ft or 1,760 yd

10 cm is about 4 in.

Time

year	365 or 366 days
year	about 52 weeks
year	12 months
month	28, 29, 30, or 31 days
week	7 days
day	24 hours
hour	60 minutes
minute	60 seconds

Money

	1¢, or $0.01	Ⓟ
	5¢, or $0.05	Ⓝ
	10¢, or $0.10	Ⓓ
	25¢, or $0.25	Ⓠ
	100¢, or $1.00	$1

Abbreviations

kilometers	km
meters	m
centimeters	cm
miles	mi
feet	ft
yards	yd
inches	in.
tons	T
pounds	lb
ounces	oz
kilograms	kg
grams	g
decimeters	dm
millimeters	mm
pints	pt
quarts	qt
gallons	gal
liters	L
milliliters	mL

Capacity

1 pint = 2 cups
1 quart = 2 pints
1 gallon = 4 quarts
1 liter = 1,000 milliliters

U.S. Traditional Addition 1

Algorithm Project 1

Write a number model to show your ballpark estimate.
Use any strategy to solve the problem.

1. Jonah had a garage sale. He earned $52
   on Saturday and $34 on Sunday. How much
   money did Jonah earn at the garage sale?

   Ballpark estimate: _____

   $ _____

Write a number model to show your ballpark estimate.
Use U.S. traditional addition to solve each problem.

2. Ballpark estimate:

   _____

```
    36
  + 23
  _____
```

3. Ballpark estimate:

   _____

```
    19
  + 76
  _____
```

4. Ballpark estimate:

   _____

```
   198
  +  46
  _____
```

5. Ballpark estimate:

   _____

```
   187
  + 123
  _____
```

U.S. Traditional Addition 2

Algorithm Project 1

Write a number model to show your ballpark estimate.
Use U.S. traditional addition to solve each problem.

1. Nia collects seashells. She had
29 seashells. She found 47 more
seashells at the beach. How many
seashells does Nia have now?

Ballpark estimate: _____

_____ seashells

---

2. Ballpark estimate:

_____

```
   41
+  38
_____
```

3. Ballpark estimate:

_____

```
   96
+  63
_____
```

---

4. Ballpark estimate:

_____

```
  187
+  78
_____
```

5. Ballpark estimate:

_____

```
  208
+ 133
_____
```

PROJECT 1 **U.S. Traditional Addition 3**

Algorithm Project 1

Write a number model to show your ballpark estimate.
Use U.S. traditional addition to solve the problem.

1. There are two second grade classes at Park
   School. One class has 34 children. The other
   has 29 children. How many children are in
   second grade at Park School?

   Ballpark estimate: _____

   _____ children

2. Write a number story for 18 + 57.
   Solve your number story.

   _____

   _____

   _____

   .   _____

   _____

Fill in the missing digits in the addition problems.

3.
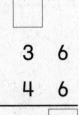

```
      □
     3 6
  +  4 6
  ───────
     8 □
```

4.
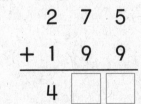

```
     □ □
    2 7 5
 +  1 9 9
 ─────────
    4 □ □
```

PROJECT 1

U.S. Traditional Addition 4

Algorithm Project 1

Write a number model to show your ballpark estimate.
Use U.S. traditional addition to solve the problem.

1. Sam had $66 in his piggy bank. He
   earned $14 babysitting this weekend.
   How much money does Sam have now?

   Ballpark estimate: _____

   $ _____

2. Write a number story for 35 + 29.
   Solve your number story.

   _____

   _____

   _____

   _____

Fill in the missing digits in the addition problems.

3.
```
      ☐
      2 9
  +   8 8
  ─────────
  1 ☐ ☐
```

4.
```
    ☐ ☐
    4 1 3
  +   8 8
  ─────────
  5 ☐ 1
```

Algorithm Project 2

Write a number model to show your ballpark estimate.
Use any strategy to solve the problem.

1. Jakob has 93 red blocks and 58 blue blocks
   in his toy box. How many more red blocks
   than blue blocks are in the toy box?

   Ballpark estimate: _____

   _____ red blocks

Write a number model to show your ballpark estimate.
Use U.S. traditional subtraction to solve each problem.

2. Ballpark estimate:

   _____

   ```
     78
   − 37
   ```

3. Ballpark estimate:

   _____

   ```
     96
   − 47
   ```

4. Ballpark estimate:

   _____

   ```
    193
   −  76
   ```

5. Ballpark estimate:

   _____

   ```
    252
   − 169
   ```

PROJECT 2
U.S. Traditional Subtraction 2

Algorithm Project 2

Write a number model to show your ballpark estimate.
Use U.S. traditional subtraction to solve each problem.

1. Rosa had 56 baseball cards.
   She gave 17 cards to her
   friend Andre. How many
   cards does she have left?

   Ballpark estimate: _____

   _____ baseball cards

---

2. Ballpark estimate:

   _____

   ```
      89
   -  71
   _____
   ```

3. Ballpark estimate:

   _____

   ```
      46
   -  28
   _____
   ```

---

4. Ballpark estimate:

   _____

   ```
     154
   -  45
   _____
   ```

5. Ballpark estimate:

   _____

   ```
     548
   - 259
   _____
   ```

U.S. Traditional Subtraction 3

Algorithm Project 2

Write a number model to show your ballpark estimate.
Use U.S. traditional subtraction to solve the problem.

1. There are 84 children who ride a
   bus to school. 65 of them are boys.
   How many girls ride a bus to school?

   Ballpark estimate: _____

   _____ girls

2. Write a number story for 79 − 68.
   Solve your number story.

   _____

   _____

   _____

   _____

   _____

Fill in the missing numbers in the subtraction problems.

3.
```
    ☐ ☐
    8̸ 4̸
  −  2 5
  ─────────
    2 ☐
```

4.
```
    ☐ ☐
  1 8̸ 7̸
  −   4 6
  ─────────
    ☐ 4 5
```

PROJECT 2

U.S. Traditional Subtraction 4

Algorithm Project 2

Write a number model to show your ballpark estimate.
Use U.S. traditional subtraction to solve the problem.

1. Jenna's book has 46 pages. Shen's book
   has 97 pages. How many more pages are
   in Shen's book?

   Ballpark estimate: _____

   _____ pages

2. Write a number story for 37 − 18.
   Solve your number story.

   _____

   _____

   _____

   _____

   _____

Fill in the missing numbers in the subtraction problems.

3.
```
   ☐ ☐
   ⁸9̸ ⁰0̸
 −  6  4
 ────────
    2  ☐
```

4.
```
      ☐ ☐
   3  ⁶7̸ ⁷8̸
 −  1  3  6
 ──────────
   ☐  ☐  7
```

Notes

Notes

Date _____ Time _____

Notes

_____

_____

_____

_____

_____

_____

_____

_____

_____

_____

_____

_____

_____

_____

Notes

LESSON 2·5

+, − Fact Triangles 1

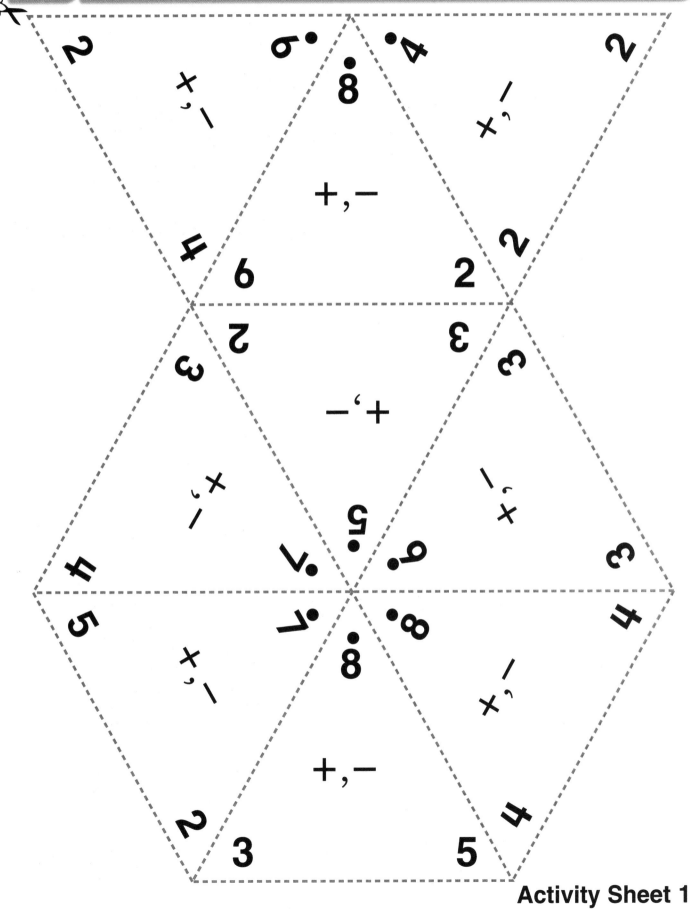

Activity Sheet 1

LESSON 2·5 **+, − Fact Triangles 2**

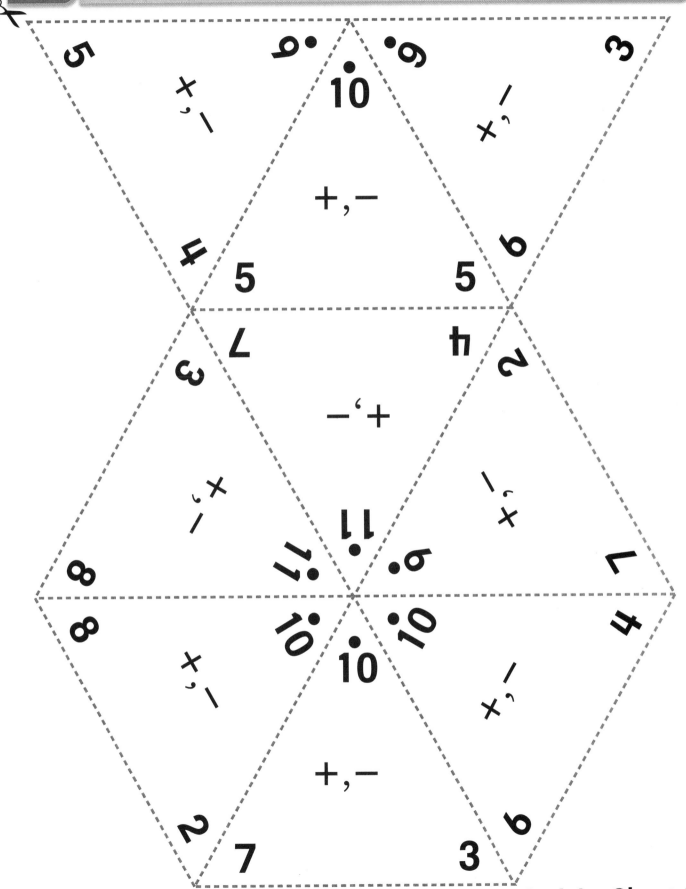

Activity Sheet 2

LESSON 2·12 +, − Fact Triangles 3

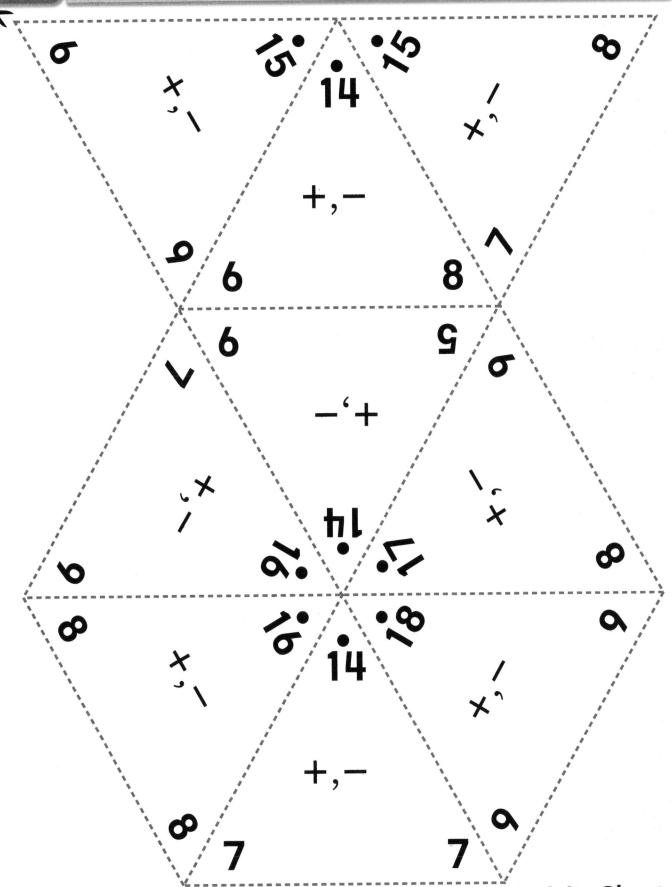

Activity Sheet 3

LESSON
2·12

+, − Fact Triangles 4

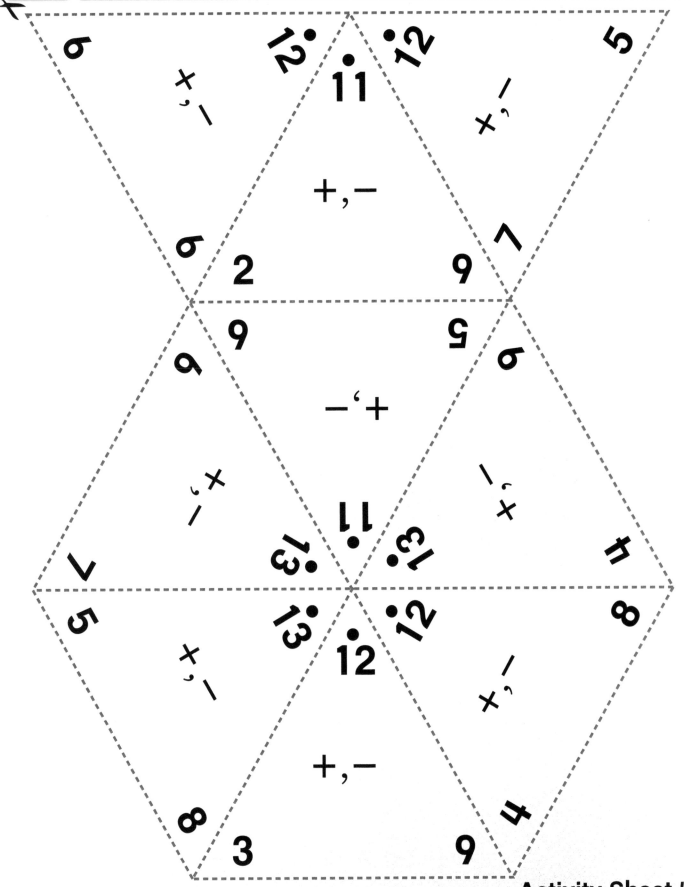

Activity Sheet 4